Aristotle's *Metaphysics* Θ 1-3

Studies in Continental Thought
John Sallis, general editor

Consulting Editors

Robert Bernasconi
Rudolf Bernet
John D. Caputo
David Carr
Edward S. Casey
Hubert L. Dreyfus
Don Ihde
David Farrell Krell
Lenore Langsdorf
Alphonso Lingis

William L. McBride
J.N. Mohanty
Mary Rawlinson
Tom Rockmore
Calvin O. Schrag
Charles E. Scott
Thomas Sheehan
Robert Sokolowski
Bruce W. Wilshire
David Wood

Martin Heidegger

Aristotle's *Metaphysics* Θ 1-3

On the Essence and Actuality of Force

Translated by
Walter Brogan and Peter Warnek

Indiana University Press
Bloomington & Indianapolis

Published in German as *Aristoteles,* Metaphysik Θ *1-3:*
Von Wesen und Wirklichkeit der Kraft.
© 1981 by Vittorio Klostermann, Frankfurt am Main.
Second edition © 1990.
© 1995 by Indiana University Press
All rights reserved

No part of this book may be reproduced or utilized in any form or by any means, electronic or mechanical, including photocopying and recording, or by any information storage and retrieval system, without permission in writing from the publisher. The Association of American University Presses' Resolution on Permissions constitutes the only exception to this prohibition.

The paper used in this publication meets the minimum requirements of American National Standard for Information Sciences—Permanence of Paper for Printed Library Materials, ANSI Z39.48-1984.

Manufactured in the United States of America

Library of Congress Cataloging-in-Publication Data
Heidegger, Martin, 1889–1976
[Aristoteles, Metaphysik, Theta 1-3. English]
Aristotle's Metaphysics [theta] 1-3: on the essence and actuality of force / Martin Heidegger; translated by Walter Brogan and Peter Warnek.
p. cm. — (Studies in Continental thought)
Includes English version of Heidegger's German translation of the Metaphysics, Book 9.
ISBN 0-253-32910-8 (alk. paper)
1. Aristotle. Metaphysics. Book 9. 2. One (The One in philosophy) I. Aristotle. Metaphysics. Book 9, 1-3. English. II. Title. III. Series.
B434.H4513 1995
110—dc20   95-834
1 2 3 4 5 00 99 98 97 96 95

Contents

Translators' Foreword ... ix

Introduction

The Aristotelian Question about the Manifold and Oneness of Being

§ 1. The question concerning δύναμις and ἐνέργεια, along with the question of the categories, belongs in the realm of the question about beings ... 1
§ 2. The manifold of the being of beings ... 8
§ 3. The equation or the differentiation of beings and being. Being as one in Parmenides ... 14
§ 4. The manifoldness and unity of being ... 21
§ 5. Oneness of being—not as genus but as analogy ... 27
§ 6. The questionableness of the analogy of being ... 34

Chapter One

Metaphysics θ 1. The Unity of the Essence of Δύναμις κατὰ Κίνησιν, Force Understood as Movement ... 40

§ 7. Considerations for the movement of the entire treatise on δύναμις and ἐνέργεια ... 40
§ 8. A subgroup of two metaphorical meanings: δύναμις with regard to the geometrical; δυνατόν and ἀδύνατον with regard to assertion ... 46
§ 9. The guiding meaning of δύναμις κατὰ κίνησιν ... 56
  a) Approaches to the phenomenon of force and a rejection of the so-called transference
  b) The apparent self-evidence of causality and the Aristotelian essential delimitation of force
§ 10. The ways of force ... 73
  a) Bearance and (prior) resistance. Effect as the being of the things of nature (Leibniz)

        b) The how which belongs to force
§ 11. The unity of the force of doing and bearing: the ontological and the ontic concepts of force and their inner adhesion   87
§ 12. Force and unforce—the carrying along with of withdrawal. The full guiding meaning   91

Chapter Two

Metaphysics Θ 2. The Division of Δύναμις κατὰ Κίνησιν for the Purpose of Elucidating Its Essence   99

§ 13. Concerning λόγος (conversance) and soul. The divisions: "conversant/without conversance" and "besouled/soulless"   99
§ 14. The extraordinary relationship of force and conversance in δύναμις μετὰ λόγου, in capability   111
        a) Capability necessarily has a realm and contraries that are in that realm
        b) The capability of producing: λόγος as innermost framework
§ 15. Δύναμις κατὰ κίνησιν as capability of the striving soul   126
§ 16. The inner divisiveness and finitude of δύναμις μετὰ λόγου   131

Chapter Three

Metaphysics Θ 3. The Actuality of Δύναμις κατὰ Κίνησιν or Capability   137

§ 17. The position and theme of this chapter and its connection to the thesis of the Megarians   137
§ 18. The beginning of Aristotle's confrontation with the Megarians   148
        a) Is the actuality of capability to be found in having or in its enactment?
        b) The conflict is grounded in the Greek understanding of actuality
§ 19. Being in practice as the actuality of capability. The phenomena of practicing and cessation   156
§ 20. The actuality of the perceptible and the actuality of the capability of perception   165
        a) The problem of the perceptible and the principle of Protagoras

b) The practicing and not-practicing of perception	
§ 21. The conclusion of the confrontation: the Megarians miss the movement of transition which belongs to a capability	177
§ 22. Ἐνέργεια κατὰ κίνησιν. The actuality of being capable is co-determined by its essence — to this essence, moreover, belongs its actuality	183
Editor's Epilogue	195
Glossary of German words	197
Glossary of Greek words	201

Translators' Foreword

This book is a translation of *Aristoteles,* Metaphysik Θ *1-3: Von Wesen und Wirklichkeit der Kraft*, which is volume 33 of Martin Heidegger's *Gesamtausgabe*. The text is based on a lecture course offered at the University of Freiburg in the summer semester of 1931. The volume presents Heidegger's translation and commentary on the first three chapters of Book Θ of Aristotle's *Metaphysics*, but Heidegger's persuasive and original interpretation of Aristotle implicates the entire corpus of Aristotle's works and leads to a rethinking of many of the central Aristotelian concepts that frame the history of Western philosophy. Heidegger's original course title, "Interpretations of Ancient Philosophy," suggests the broader context in which he situates Aristotle's investigation of δύναμις and ἐνέργεια in *Metaphysics* Θ. In this treatise, Aristotle brings *the* question of Greek philosophy, the question of being, to its sharpest formulation.

These lectures, along with Heidegger's other courses on Aristotle, offer an immense contribution to the advancement of Aristotle scholarship and have had a wide influence on the development of Aristotle studies in Europe. It is telling that so many of Heidegger's students during the early years of his teaching have themselves in turn offered major contributions to Aristotle scholarship. Often, according to their own testimonies, these works have been presented under the direct influence and guidance of Heidegger's lectures. Thus Helene Weiss, in her work on Aristotle, writes: "I have freely made use of the results of Heidegger's Aristotle interpretation which he delivered in lectures and seminars." In their works on Aristotle, Pierre Aubenque, Jean Beaufret, Walter Bröcker, Ernst Tügendhat, and Fridolin Wiplinger, among others, all equally acknowledge their indebtedness to Heidegger's revolutionary interpretations. Evidence of the continuing impact of Heidegger's Aristotle interpretation on contemporary European Aristotle scholarship can be seen, for example, in the work of Rémi Brague.

The lectures on Aristotle herein presented further confirm the widely acknowledged influence of Aristotle on Heidegger's own original thought. Heidegger says in *Being and Time:* "My task is the same task that provided the impetus for the research of Plato and Aristotle, only to subside from then on as an actual theme for investigation." And he comments in "My Way to Phenomenology": "Of course I could not immediately see what decisive consequences my renewed preoccupation with Aristotle was to have." On another occasion, Heidegger calls his study of Brentano's work on Aristotle "the ceaseless impetus for the treatise *Sein und Zeit.*" Many of Heidegger's students, notably Walter Biemel, Hans-Georg Gadamer, and William Richardson, have long insisted that Heidegger's studies of Aristotle were of paramount importance for the development of his own phenomenology. With the publication of this volume of the "Collected Works," especially the material in the second chapter, and the publication in 1992 of volume 19, which contains Heidegger's analysis of Book VI of Aristotle's *Nicomachean Ethics*, the long-intimated connection between Heidegger's work on Aristotle and his Dasein analysis can begin to be seen. Indeed, much of Heidegger's teaching prior to the publication of *Being and Time* was related to his work on Aristotle.

The publication in 1993 of volume 22 of the "Collected Works," Heidegger's 1926 course on "The Basic Concepts of Ancient Philosophy," further contributes to the scholarly process that has been undertaken to make these courses on Aristotle and other previously unpublished works of Heidegger available. And the recent recovery of the formative 1922 manuscript on Aristotle, referred to as a first draft of *Being and Time*, the writing that Heidegger had sent to Marburg and Göttingen in support of his nomination for a position at these institutions, helps to further our understanding of the important link between Heidegger's early work on Aristotle and the development of his own method of phenomenology. Although delivered in Freiburg at a later period, the lectures that constitute this volume on *Metaphys-*

ics Θ 1-3 are clearly the fruit of his earlier Marburg lectures on Aristotle. The entire discussion of Book Θ 1-3 is linked to the discussion of truth which takes place in chapter ten of Book Θ. Thus, this lecture course, though offered later, provides the necessary background for Heidegger's important discussion of *Metaphysics* Θ 10 in the Marburg course of 1925/26: *Logik: Die Frage nach der Wahrheit*, which is published as volume 21 of the "Collected Works." We can see the extent to which Heidegger's reading of Aristotle is at variance with other contemporary scholarship in his disagreement with Werner Jaeger about Aristotle's treatment of truth in *Metaphysics* θ 10. According to Jaeger, the tenth chapter on truth must be spurious, a later imposition which does not fit in with Aristotle's thought. In contrast, Heidegger says in this volume: "With this chapter, the treatise reaches its proper end; indeed, the whole of Aristotle's philosophy attains its 'highest point.'"

Translating is inherently an ambiguous task, and Martin Heidegger's point that every translation is an interpretation cannot be denied. At the same time, the translator aims as much as possible to present the original text accurately and without interference. This particular text poses its own special difficulty for a translator. Much of the German text is itself a line-by-line translation of Aristotle's Greek. Heidegger's translations are often literal and nuanced, rendering the Greek in multiple alternatives, sacrificing fluency so as to allow the German to express Aristotle's philosophical thought. Translating into English a German text that stays so close to the Greek without resorting to artificially awkward English style presents a challenge. We have attempted to combine accurate and literal renditions with an attentiveness to the readability of the text.

The glossary at the back of the book lists some of the more important translation choices we have made. We call attention here only to several particularly crucial decisions. The many meanings of λόγος for Aristotle and the essential interconnection of these meanings is one of the central issues of Heidegger's lectures. He offers a number of translations and interpretations of λόγος, but in chapter two finally settles on *Kundschaft*, a word that brings together these multiple meanings. We have translated *Kundschaft* as "conversance." To be conversant is to have expert knowledge of something, to be familiar with it, and also to be able to speak of it with a sense of its surroundings. In light of its importance in this text, it is surprising that the word *Kundschaft* as a translation of λόγος remains peculiar to this text.

Among the most difficult passages to translate were those in the Introduction that deal with the meaning of being in Greek (τὸ ὄν/τὸ εἶναι). The German variations (*das Sein, die Seiendheit, das Seiende, ein Seiendes, das Seiend*) and their philosophically important interplay have no corresponding terms in English. We have chosen not to use the capital *B* for "being" or alternate words, such as "entities," for "beings." Instead we have included the German in brackets when it was helpful for the reader to follow Heidegger's discussion. Though Heidegger employs a vast array of related words to translate δύναμις in various contexts, most often he uses the word *Kraft*. Heidegger himself points out that *Kraft* has many related meanings in German. Whenever possible, we have followed the common practice of translating *Kraft* as "force." However, when the context dictated, we have also used other words, most notably "power," to translate *Kraft*. We struggled to come up with a better translation for *Vorhandensein*, but we have rendered it imperfectly as "being-present." Finally, the two cognate words *Vollzug* and *Entzug* have been translated as "enactment" and "withdrawal." Unfortunately, these translations do not preserve the philosophically important interconnection that can be noted more easily in German because of the common suffix -*zug*.

We appreciate the very helpful suggestions made by John Protevi, who critiqued our translation. We also appreciate the support of our colleagues and students. We would especially like to thank John Sallis for his patient encouragement and confidence. For both of us, he is a mentor who exemplifies how to read the texts of Heidegger. We are enormously indebted to Elaine Brogan and Stephanie Jocums. They have not only helped prepare the manuscript; they have been an integral and generous part of this entire project. We are thankful for the hospitality of Peter Kessler, who provided us an opportunity to work on this translation for several weeks in a villa on Lake Zürich whose beauty nourished the intensity and excitement of two friends working together in philosophy. Finally, we are grateful to the National Endowment for the Humanities for their partial support of this project.

Walter Brogan and Peter Warnek

Aristotle's *Metaphysics* Θ 1-3

The inner will of this course can be characterized by a word from Nietzsche:

> Perhaps some centuries later one will judge that all German philosophy finds its authentic worth in that it is a gradual recovery of the soil of antiquity, and that each claim to "originality" sounds trite and laughable in relation to the higher claim of the Germans to have reestablished the apparently broken link with the Greeks, up until now the highest type of "human being."
>
> —*The Will to Power*, Aph. 419

Introduction

The Aristotelian Question about
the Manifold and Oneness of Being

*§ 1. The question concerning δύναμις and ἐνέργεια, along with
the question of the categories, belongs in the realm of
the question about beings*

This course confronts the task of interpreting philosophically a philosophical treatise of Greek philosophy. The treatise has come down to us as Book IX of Aristotle's *Metaphysics*. It is a self-contained unit, divided into ten chapters, whose object of inquiry is δύναμις and ἐνέργεια. These words are translated into Latin as *potentia* and *actus* and into German as *Vermögen* [capability] and *Verwirklichung* [actualization] or as *Möglichkeit* [potentiality] and *Wirklichkeit* [actuality].

What is being sought in this inquiry into δύναμις and ἐνέργεια? What prompts this investigation of "potentiality and actuality"? In what encompassing realm of questioning does this treatise of Aristotle's belong?

The answer to these questions is not far out of reach. We need only observe the context of the inquiry: Aristotle's *Metaphysics*. Thus the treatise on δύναμις and ἐνέργεια is metaphysical. This account is reasonable enough and accurate; but for just this reason it tells us absolutely nothing. Do we really know what this thing is that we so commonly call "metaphysics"? We do not. Nowadays the word bewitches us like a magical incantation with its suggestion of profundity and its promise of salvation. But the information that this treatise by Aristotle is metaphysical not only says nothing; it is downright misleading. And this is true not only today; it has been true for the last two thousand years. Aristotle never had in his possession what later came to be understood by the word or the concept "metaphysics." Nor did he ever seek anything like the "metaphysics" that has for ages been attributed to him.

If we do not let ourselves be swayed by the tradition and resist being talked into anything, and if we therefore reject the readily available information that the treatise is "metaphysical," what then? How else are we to locate the realm of questioning in which the treatise belongs? Or should we leave the matter open and undetermined? In which case, our attempt to enter into Aristotle's inquiry, and thus to inquire along with it, would be without direction or guidance for some time. Before we begin, we need to clarify the aim of this treatise, as well as its sequence and the scope of its point of departure. In what realm of questioning, then, does the treatise belong? The text itself provides the answer in its first few lines:

Περὶ μὲν οὖν τοῦ πρώτως ὄντος καὶ πρὸς ὃ πᾶσαι αἱ ἄλλαι κατηγορίαι τοῦ ὄντος ἀναφέρονται εἴρηται, περὶ τῆς οὐσίας. κατὰ γὰρ τὸν τῆς οὐσίας λόγον λέγεται τἆλλα ὄντα, τό τε ποσὸν καὶ τὸ ποιὸν καὶ τἆλλα τὰ οὕτω λεγόμενα· πάντα γὰρ ἕξει τὸν τῆς οὐσίας λόγον, ὥσπερ εἴπομεν ἐν τοῖς πρώτοις λόγοις (Θ 1, 1045b27–32)

"We have thus dealt with beings in the primary sense, and that means, with that to which all the other categories of beings are referred back, οὐσία. The other beings"—please note: τὸ ὄν: being [*das Seiend*] (participial!)—"the other beings (those not understood as οὐσία) are said with regard to what is said when saying οὐσία, the how much as well as the how constituted and the others that are said in this manner. For everything that is (the other categories besides οὐσία) must in and of itself have the saying of οὐσία, as stated in the previous discussion (about οὐσία)." (Regarding πρώτως: the sustaining and leading fundamental meaning, see below, p. 34ff.).

The first sentence establishes that the categories, and in fact the first category, have already been dealt with in another treatise. The second sentence characterizes the manner of the relation back and forth of the other categories to the first. Three times in this sentence we find: λόγος, λέγεται, λεγόμενον. The relation back and forth of the other categories occurs as λέγειν in the λόγος.[1]

Λέγειν means "to glean" [*lesen*], that is, to harvest, to gather, to

---

1. Concerning logos, see the beginning of the Sophist course in the winter semester of 1924–25: insufficient.

§ 1. *Question concerning* δύναμις *and* ἐνέργεια

add one to the other, to include and connect one with the other. Such laying together is a laying open [*Dar-legen*] and laying forth [*Vor-legen*] (a placing alongside and presenting) [*ein Bei- und Dar-stellen*]: *a making something accessible in a gathered and unified way*. And since such a gathering laying open and laying forth occurs above all in recounting and speaking (in trans-mitting and com-municating to others), λόγος comes to mean discourse that combines and explains. Λόγος as laying open is then at the same time evidence [*Be-legen*]; finally it comes to mean laying something out in an interpretation [*Aus-legen*], ἑρμηνεία. The meaning of λόγος as relation (unified gathering, coherence, rule) is therefore "prior" to its meaning as discourse (see below, p. 103). Asking how λόγος also came to have the meaning of "relation" is therefore backwards; the order of things is quite the reverse.

The gathering and explaining of discourse makes things accessible and manifest. Heraclitus, for example, says this in Fragment 93: ὁ ἄναξ, οὗ τὸ μαντεῖόν ἐστι τὸ ἐν Δελφοῖς, οὔτε λέγει οὔτε κρύπτει . . . "The lord whose oracle is at Delphi neither speaks out nor conceals." On the basis of this stark contrast between λέγειν and κρύπτειν, to conceal, it is made simply and emphatically clear that λέγειν, as distinguished from concealing, is revealing, making manifest. Plato definitively makes the point in the *Sophist* when, toward the end of the dialogue, he understands the inner province of λόγος as δηλοῦν, making manifest. λόγος as discourse is the combining and making manifest in the saying, the unveiling assertion of something about something.

The relation back and forth of the other categories to the first category, which Aristotle discusses, occurs in λόγος. Accordingly, when we say succinctly that this relation back and forth of the categories to the first is "logical," this means only that this relation is founded in λόγος—in the elucidated sense of the word. We should once and for all steer clear of all the traditional and usual ideas about the "logical" and "logic," assuming that we are thinking of anything definite or truly fundamental with these words "logical" and "logic."

And if in fact what we call categories are not just found in λόγος and are not just used in assertions, but *by virtue of their essence have their home in* assertions, then it becomes clear why the categories are precisely called "categories."

Κατηγορεῖν means to accuse, to charge—thus to begin with not just any assertion, but one that is emphatic and accentuated. It is to say something to someone's face, to say that one is so and so and that this is one's situation. Applied to things and to beings in general, it is the kind of saying which says emphatically *what a being properly is* and how it is; κατηγορία is therefore anything said or sayable in this way. If the categories have their home in λόγος, then this means that in every assertion whatsoever of something about something, there is that exceptional saying wherein the being as it were is rightly indicted for being what it is. Aristotle sometimes also uses κατηγορία in the broad, attenuated sense of what is said in verbal transactions, what is simply asserted; or better (see *Physics* B 1, 192b17), the simple claim [*An-spruch*], that which one literally has given one's word to—the name, the word, and the relationship to the thing. What Aristotle calls "category" in the stated sense, however, is that saying which is involved in every assertion in a preeminent way (even when this is not expressed).

It is convenient and therefore popular, particularly in giving an account of ancient philosophy, to appeal to later and more recent doctrines to aid in understanding. On the question of the Aristotelian categories, one usually consults Kant. And in point of fact, he also derived the categories "logically" from the Table of Judgments, from the modes of assertion. But "logical" for Kant and "logical" for Aristotle have different meanings. Not only that. The comparison above all overlooks a fundamental character of the categories as Aristotle understands them. This fundamental character of the categories is expressly stated in the passage we are considering: κατηγορίαι τοῦ ὄντος, "*categories of beings.*" What does this mean? Does it mean categories that refer to beings as to their "object" (*genitivus objectivus*) or categories that belong to beings as to a subject (*genitivus subjectivus*)? Or are both meant? Or neither? We shall have to leave this question open.

In any case, the usual representation of the categories as "forms of thought," as some sort of encasements into which we stuff beings, is thereby already repudiated for having mistaken the facts. All the more so considering that Aristotle in our passage even calls the categories

§ 1. Question concerning δύναμις and ἐνέργεια

simply τὰ ὄντα, "*beings*" [*die Seienden*], that which absolutely belongs to beings.

Yet earlier, in our interpretation of Aristotle's second sentence in the above passage, we said that the categories have their home in λόγος. But λόγος, assertion, is assertion about beings, not the beings themselves. So we have a dual claim: the categories belong to λόγος, and the categories are the beings themselves. How do these go together? We do not have the answer. From now on let us remember that the question of the essence of the categories leads into obscurity.

(The essence of the categories is rooted in λόγος as a gathering and making manifest. Does this connection of oneness and truth signify being? At that place in Parmenides where the first saying of being occurs, the character of presence is ἕν [compare p. 19 below]. Notice the interconnection of ὄν as οὐσία, παρ- and συνουσία, and ἕν as together with, and λόγος as gatheredness, assemblage, consolidation; and in this context the "copula," the "is.")

The third sentence we cited from Aristotle in the above passage further determines in what sense οὐσία is first among the categories. "For everything that is must of and in itself have the saying of οὐσία." For example, ποιόν, the being so constituted [*Beschaffensein*]. Taken alone, there is no such thing. We do not understand being so constituted in its most proper meaning unless we comprehend as well the being so constituted of something. This reference—"of something"—is part of the very makeup of the categories. The other categories are not only incidentally and subsequently connected with the first category by means of assertions, as though they could mean something independently; rather, they are always, in accord with their essence, *co-saying* the οὐσία. And to the extent that the categories are beings, they are co-being with οὐσία. This is already said beforehand and being beforehand. It is the first category, and that also means the first being: τὸ πρώτως ὄν.

So much for a rough exposition of the beginning of our treatise. At the moment we are only trying to discover from these introductory sentences the realm of questioning in which the treatise itself is located. Do the sentences just discussed say anything about this? On the contrary, this is a mere summary of what was discussed in another treatise.

An inquiry about the categories, in fact the first category, has come down to us as Book VII (Z) of the *Metaphysics*. But perhaps Aristotle recalls precisely this treatise on οὐσία (which is also self-contained) in order to suggest that the following treatise also pertains to the realm of the question of the categories; δύναμις and ἐνέργεια, which are to be dealt with now, would then be two additional categories that receive special examination. This viewpoint suggests itself when we consider the later and now common conception of δύναμις and ἐνέργεια, possibility and actuality. For Kant above all, and since Kant, "possibility" and "actuality," along with "necessity," belong among the categories; in fact, they form the group of categories called "modality." They are, as we say in short, modalities. But we do not find δύναμις and ἐνέργεια in any of Aristotle's enumerations of the categories. *For Aristotle, the question of* δύναμις *and* ἐνέργεια, *possibility and actuality, is not a category question.* This shall be maintained unequivocally, despite all conventional interpretations to the contrary. And this clarification (though admittedly it is once again only negative) is the primary presupposition for understanding the entire treatise.[2]

But then where does this inquiry belong, if not in the framework of the category question? This is stated very clearly in the following sentence (Θ 1, 1045b32–35):

ἐπεὶ δὲ λέγεται τὸ ὂν τὸ μὲν τὸ τί ἢ ποιὸν ἢ ποσόν, τὸ δὲ κατὰ δύναμιν καὶ ἐντελέχειαν καὶ κατὰ τὸ ἔργον, διορίσωμεν καὶ περὶ δυνάμεως καὶ ἐντελεχείας. "But since beings are said on the one hand (τὸ μέν) as what being, or being so and so or so much being (in short, in the sense of the categories), and on the other hand (τὸ δέ) in regard to δύναμις and ἐντελέχεια and ἔργον, so shall we also undertake a conceptually sharp elucidation of δύναμις and ἐντελέχεια."

Here we see clearly that the question concerning δύναμις and ἐντελέχεια is also a question about beings as such, but one that aims in another direction. Thus like the category question it revolves

---

2. This would also be an appropriate place to discuss τὸ ὑπάρχειν, τὸ ἐξ ἀνάγκης, and τὸ ἐνδέχεσθαι ὑπάρχειν, *Analytica priora*, A 2, 25a1f.; compare *De interpretatione*, chap. 12f.

§ 1. Question concerning δύναμις and ἐνέργεια

around *the general realm of the question of beings*, which is the only question that fundamentally interests Aristotle.

Hence τὸ ὄν, beings themselves, according to their essence, must permit the one discussion (in the sense of the categories) as well as the other (in the sense of δύναμις and ἐνέργεια)—indeed, not only permit but perhaps require. The discussion of δύναμις and ἐνέργεια is therefore a questioning about beings aimed in a specific direction and differentiated from the question of the categories. Yet the questioning of what beings are insofar as they are beings—τί τὸ ὄν ᾗ ὄν—this questioning, as Aristotle often points out (for example, *Met.* Γ 1 and 2 and Ε 1),[3] is the most proper form of philosophizing (πρώτη φιλοσοφία). Since this questioning is directed at the πρώτως ὄν (οὐσία), do not δύναμις and ἐνέργεια then belong to οὐσία? Or is this belonging precisely the question? Inasmuch as this question has not been posed, much less answered, this indicates that we have not come to terms with the question of being.

To inquire into δύναμις and ἐνέργεια, as Aristotle proposes to do in our treatise, is genuine philosophizing. Accordingly, if we ourselves have eyes to see and ears to hear, if we have the right disposition and are truly willing, then, if we are successful, we will learn from the interpretation of the treatise what philosophizing is. We will in this way gain an experience with philosophizing and perhaps become more experienced in it ourselves.

The treatise on δύναμις and ἐνέργεια is *one* of the ways of questioning about beings as such. Aristotle does not say any more here. Rather, after the sentences we just read (1045b35ff.), Aristotle proceeds immediately to more closely delimit his topic and to delineate the course of the whole inquiry.

Aristotle does not say any more. But it is enough, more than enough, for us—we who come from afar and from outside and who no longer have the ground that sustains this inquiry. It is more than enough for us who have been trained in indiscriminate philosophical scholarship and who conceal our philosophical impotence in clever

---

3. See the passages in other of his writings where πρώτη φιλοσοφία is mentioned in Bonitz's *Commentary*, prologue, p. 3f.

industry. What we need is a brief pause to reflect on what is said here in the announcement of the treatise's realm of questioning. We want to see more clearly what Aristotle asks about and how this questioning is worked through.

§ 2. The manifold of the being of beings

Beings are said and addressed sometimes in the mode of categories, and sometimes in that of δύναμις—ἐνέργεια; thus in a dual way, διχῶς, not μοναχῶς, not in a single and simple way. What is the origin of this distinction? What is the justification for this twofold deployment in the address and saying of being? Aristotle offers no explanation or reason for this, neither here nor elsewhere. He does not even so much as raise the question. This differentiation of the ὄν is simply put forth. It is somewhat like when we say that mammals and birds are included in the class of animals. Τὸ ὄν λέγεται τὸ μέν—τὸ δέ. Why are beings deployed in such a twofold way? Is it because of the beings themselves that we have to give this dual account of beings? Or are we humans the reason for it? Or is this solely due neither to beings themselves nor to us humans? But then to what is it due?

As soon as we probe, albeit in but a general way, into the realm of the question of beings, we find ourselves once again in obscurity. But our perplexity increases further in that we are not permitted to remain content with the mere either-or: beings are said either in the manner of the categories or in the manner of δύναμις and ἐνέργεια. Aristotle himself apprises us of yet another way in the beginning of the tenth and final chapter of this very treatise.[4] (With this chapter, the treatise reaches its proper end; indeed, the whole of Aristotle's philosophy attains its "highest point.") Chapter ten begins:

ἐπεὶ δὲ τὸ ὄν λέγεται καὶ τὸ μὴ ὄν τὸ μὲν κατὰ τὰ σχήματα τῶν

---

4. See the interpretation of *Metaphysics* Θ 10 found in the 1930 summer semester course.

§ 2. Manifold of the being of beings

κατηγοριῶν, τὸ δὲ κατὰ δύναμιν ἢ ἐνέργειαν τούτων ἢ τἀναντία, τὸ δὲ κυριώτατα ὂν ἀληθὲς ἢ ψεῦδος . . . (1051a43–b1)

"Since the being [*das Seiende*] and non-being are said on the one hand in accordance with the forms of categories, and on the other hand in accordance with the potentiality and actuality of these or of their contrary (in short: in accordance with δύναμις and ἐνέργεια), but the most authoritative being [*das Seiende*] is true and untrue being . . ."

Without going into detail about all that the beginning of this chapter offers that is new in comparison to that of chapter one, we can see one thing clearly: the passage again has to do with the folding [*Faltung*] of the being. The being with respect to the categories and the being with respect to δύναμις and ἐνέργεια are again mentioned in the same order. But then a third is added, namely, the *true-or-untrue being*. The being is not dually (διχῶς) folded but triply (τριχῶς) folded. Thus the question of the unfolding of the being and the origin of this unfolding becomes noticeably more complicated; the inner interconnection of the three foldings becomes more impenetrable.

And if we check carefully to see if and how Aristotle himself organizes the questioning concerning beings, then it will become apparent that, precisely in the passage where he specifically proposes to identify and fully review the folding of beings, Aristotle lists not a threefold folding (τριχῶς) but a fourfold (τετραχῶς). He says in the beginning of chapter two of the treatise we know as Book VI (E) of the *Metaphysics*:

ἀλλ' ἐπεὶ τὸ ὂν τὸ ἁπλῶς λεγόμενον λέγεται πολλαχῶς, ὧν ἓν μὲν ἦν τὸ κατὰ συμβεβηκός, ἕτερον δὲ τὸ ὡς ἀληθές, καὶ τὸ μὴ ὂν ὡς ψεῦδος, παρὰ ταῦτα δ'ἐστὶ τὰ σχήματα τῆς κατηγορίας, οἷον τὸ μὲν τί, τὸ δὲ ποιόν, τὸ δὲ ποσόν, τὸ δὲ ποῦ, τὸ δὲ ποτέ, καὶ εἴ τι ἄλλο σημαίνει τὸν τρόπον τοῦτον· ἔτι παρὰ ταῦτα πάντα τὸ δυνάμει καὶ ἐνεργείᾳ· ἐπεὶ δὴ πολλαχῶς λέγεται τὸ ὄν, πρῶτον . . . (1026a33–b2)

"But since beings that are addressed purely and simply are said in various ways, one of which is the being [*das Seiende*] with respect to being co-present, another is the being as true and the non-being as untrue; and in addition to these, the being according to the forms of the categories—the what, the how, the how much, the where, the when, and others that signify the being in the same way; besides all these, the being in the sense of δύναμις and ἐνέργεια; since, therefore, beings are addressed in various ways, it is necessary . . ."

In addition to the three previously listed foldings, Aristotle here gives a fourth, which he in fact mentions first—the ὂν κατὰ συμβεβηκός—the *accidental being*. Συμβεβηκός: accidental [*zufällig*]; this is a remarkable translation. The translation actually follows the Greek literally; yet it does not hit upon the true Aristotelian meaning. The accidental is indeed a συμβεβηκός, but not every συμβεβηκός is accidental.

The sequence of the list is different here. But this is at first of no consequence. More important is the way that Aristotle introduces the four foldings of beings, both here and in the other passages: ἓν μέν, ἕτερον δέ, παρὰ ταῦτα δέ, ἔτι παρὰ ταῦτα πάντα—on the one hand, on the other hand, in addition to, besides. It is a simple serial juxtaposition without any consideration of their structure or connection, much less their justification. Only one thing is said: τὸ ὂν λέγεται πολλαχῶς—*beings are said in many ways;* in fact, in four ways.

This sentence, τὸ ὂν λέγεται πολλαχῶς, is a constant refrain in Aristotle. But it is not just a formula. Rather, in this short sentence, Aristotle formulates the wholly fundamental and new position that he worked out in philosophy in relation to all of his predecessors, including Plato; not in the sense of a system but in the sense of a task.

*

Our task is the interpretation of *Metaphysics* Θ, the inquiry concerning δύναμις and ἐνέργεια. In order that we may be in a position to join in the inquiry, and not just for instructional reasons, an indispensable preparation is required, namely the designation and delimitation of the realm of questioning in which Aristotle is here inquiring. The treatise itself throws some light on this matter. It begins by referring to another treatise which dealt with the categories. The categories are founded in λόγος—they are τὰ ὄντα, beings themselves. As always, the question of the categories is a question of the ὄν. Τὸ

§ 2. Manifold of the being of beings

ὄν, however, is also said κατὰ δύναμιν καὶ ἐνέργειαν. Therefore an investigation of this is to come next. The realm is τὸ ὄν, beings. At first, beings are distinguished in two ways (διχῶς), later in three ways (τριχῶς), and finally in four ways (τετραχῶς). Again and again Aristotle says: τὸ ὂν λέγεται πολλαχῶς. But the juxtaposition is retained.

The programmatic assertion of the fourfold folding of beings (E 2) is especially important in that it adds a more precise determination to the ὄν here being articulated: Τὸ ὄν—τὸ ἁπλῶς λεγόμενον, the being, that is, that which is addressed purely and simply in itself; that is to say, the being taken purely as itself, precisely as being; the ὂν ᾗ ὄν, the being inasmuch as it is a being.

What do we have in mind when we address beings as beings? What can and must we say of beings when they are considered solely and specifically as beings? We say: The being is. What makes beings beings is *being*. Thus when Aristotle speaks of the variety of the folding of beings as beings, he means the manifold of the folding of the being of beings. Being unfolds itself.

Chapter seven of Book V (Δ) of the *Metaphysics* makes it absolutely clear that Aristotle is referring to the being of beings. Book V is by no means constituted of a single investigative treatise. It is rather a compilation of the various meanings of some basic concepts of philosophy. Chapter seven enumerates the different meanings of ὄν, namely the four that we just became acquainted with in E 2. Here in Δ 7, they are again presented in a different sequence.

The chapter begins (1017a7): τὸ ὂν λέγεται . . . In introducing ὄν according to the forms of the categories, Aristotle says: ὁσαχῶς γὰρ λέγεται [τὰ σχήματα τῆς κατγορίας, H.] τοσαυταχῶς τὸ εἶναι σημαίνει. "For as the forms of the category are said in various ways, so being has various meanings." Incidentally, ἡ κατηγορία is an indication: the singular here signifies the preeminent saying of the being [*das Seiende*] in every individual assertion about this or that being. The category: the saying of being in the assertion (λόγος) of beings. Τὸ εἶναι is found instead of τὸ ὄν in the cited sentence; that is, τὸ ὄν is understood in the sense of τὸ ὂν ᾗ ὄν. And it is the same with the remaining modes of ὄν: ἔτι τὸ εἶναι σημαίνει καὶ τὸ ἔστιν ὅτι ἀληθές (a31, a passage of great importance that we cannot go into at this time). "Furthermore, being signifies 'is' in the sense of 'it is true.'" Just as we too say something *is* so—in emphasizing the "is," we mean to say: it is in truth so. Here then the concern is with the being of being true. Finally, the ὄν as δύναμις and ἐνέργεια is introduced (a35f.): ἔτι τὸ εἶναι σημαίνει καὶ τὸ ὄν τὸ μὲν δυνάμει ['ῥητόν], τὸ δ'ἐντελεχείᾳ; "Furthermore, being also means the being δυνάμει as well as ἐντελεχείᾳ." Τὸ εἶναι σημαίνει τὸ ὄν: being means the being [*das Seiende*] (actually being [*Seiend*] and not beings). Being (εἶναι) means nothing other than the being (ὄν) insofar as the being is this and nothing other.

The realm of questioning of our treatise is the ὄν ᾗ ὄν: beings as beings; but now this means being. And what is being asked about is a way of being that folds itself in four foldings that are simply listed in a row. Τὸ ὄν λέγεται πολλαχῶς means: τὸ εἶναι (τοῦ ὄντος) λέγεται πολλαχῶς. The πολλαχῶς ascribed to ὄν and εἶναι refers in most cases to the four ways of being mentioned above, even when at times only two or three of these are listed: πολλαχῶς = τετραχῶς.

However, ὄν πολλαχῶς λεγόμενον also has a narrower meaning. In this case, it does not refer to the aforementioned four ways, but to one of them, one which again and again assumes a certain priority: τὸ ὄν κατὰ τὰ σχήματα τῆς κατγορίας—being in the sense of the category. This ὄν, that is, εἶναι in this sense, is not only one among the πολλαχῶς of four, it is in itself a πολλαχῶς λεγόμενον, namely, in as many ways as there are categories. Compare a23f.: ὁσαχῶς γὰρ λέγεται [ἡ κατηγορία, H.] τοσαυταχῶς τὸ εἶναι σημαίνει. Thus, this ὄν is itself a πολλαχῶς λεγόμενον because here the λέγειν is the utterly exceptional λέγειν of the κατηγορία which already prevails in any λόγος whatsoever.

The treatise of the *Metaphysics* that discusses the first category, οὐσία, and which is one of the cornerstones of Aristotle's philosophy, is in accord with all of this when it begins with the simple guiding proposition: τὸ ὄν λέγεται πολλαχῶς (Z 1, 1028a10). What follows next in the text that has been handed down, namely καθάπερ διειλόμεθα . . . ποσαχῶς, could not have come from Aristotle but was inserted later by those who attempted to paste together the indi-

§ 2. Manifold of the being of beings

vidual treatises of Aristotle into a so-called work. The same is the case with the final sentence of the preceding book.[5] Thus, Book Z, with some justification, came to be classified in the realm of the fourfold questioning of ὄν. But the πολλαχῶς with which this book begins refers to something else. What that is is clearly stated in the next sentence (a11ff.): σημαίναι γὰρ τὸ μὲν τί ἐστι καὶ τόδε τι, τὸ δὲ ὅτι ποιὸν ἢ ποσὸν ἢ τῶν ἄλλων ἕκαστον τῶν οὕτω κατηγορουμένων. "For the being means on the one hand what-being and this-being, and on the other hand it means that it (the one just named) is of such and such a kind or so and so much, etc." The τὸ μέν—τὸ δέ is arranged in such a way that οὐσία is alone on one side and the rest are on the other side. The πολλαχῶς here refers to a multiplicity within "the category"; and, in fact, this manifoldness has a certain order and arrangement, namely the one with which we are already familiar—the ordering of all the remaining categories up to the first (see the beginning of *Met.* Θ 1). The next sentence (Z 1, 1028a13ff.) points this out: τοσαυταχῶς δὲ λεγομένου τοῦ ὄντος φανερὸν ὅτι τούτων πρῶτον ὂν τὸ τί ἐστιν, ὅπερ σημαίνει τὴν οὐσίαν. "As variously as the being may be said here (thus it is not a matter of a confused and arbitrary manifoldness, but) it is apparent from this that first being is what-being, which means οὐσία." According to the above, οὐσία therefore means: τὸ ἁπλῶς πρώτως λεγόμενον—but ἁπλῶς (simple) also in contrast to πολλαχῶς in the wider sense.

Thus we have the πολλαχῶς of the various categories within the wider πολλαχῶς (τετραχῶς). The πολλαχῶς as such is itself a διχῶς λεγόμενον; it is said doubly. The connection between these two must be seen clearly, not only in order to get to know Aristotle's use of language but so that the Aristotelian question of ὄν can be conceived philosophically.

---

5. Compare the remarks of Christ on this passage (missed by Bonitz and Schwegler); of course, Christ does not consider, nor was it his task to consider, the basis of this differentiation: the narrower meaning of πολλαχῶς.

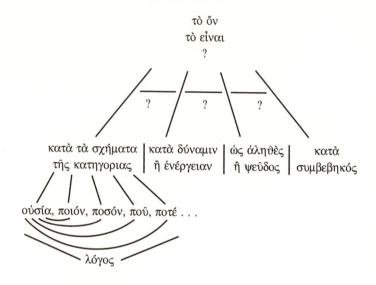

§ 3. *The equation or the differentiation of beings and being. Being as one in Parmenides*

We have seen that the question τὶ τὸ ὄν—what are beings?—is the question τί τὸ εἶναι, what is being? How can Aristotle equate τὸ ὄν and τὸ εἶναι? Why say the question is about beings (ὄν) when it is about being (εἶναι)?

Even today, we still commonly make this equation, although more with a sense of a hopeless confusion. Thus we speak often in philosophy of being and mean beings. On the other hand, we say beings and mean being. Basically we comprehend neither the one nor the other. And yet we do understand something when we say this, even though everything dissolves into thin air when we attempt to grasp it. For example, this thing here, this piece of chalk, is a being, it "is"; we say this of the chalk because it, as it were, says this to us in advance. In the same way, my speaking now and your listening and paying attention are [ways of] being. We experience and grasp beings constantly and with ease. But "being"? In a certain sense we understand this also but do not comprehend it. How then are we to distinguish

§ 3. Equation or differentiation of beings

the two, beings and being, or even understand the relevant inner relationship between the two, if all this has not been assimilated or for that matter even expressly questioned?

Who would deny that philosophers have been discussing beings and being for a long time? This equation of τὸ ὄν and τὸ εἶναι, beings and being, has a venerable tradition. We already encounter it in Parmenides at the decisive beginning of Western philosophy: τὰ ἐόντα and ἔμμεναι(an older Lesbian form).[1] But this equation is at the same time a hidden and not understood difference that does not come into its own.

Is the time-honored character of this equation in itself sufficient indication of the lucidity of what is equated, or indeed an argument for the legitimacy of speaking this way? The fact that we ordinarily speak this way cannot be faulted. That we do so when expressly inquiring about beings is, however, to say the least, peculiar. Or is the question really not about beings at all? Yet Aristotle and the philosophers preceding him certainly do ask expressly about this. Or is this simple questioning not enough? Is it only an initial approach that subsequently comes to a standstill? Being and beings—still to make a distinction here or even to want to raise a question, is this not unwarranted, futile quibbling? Beings and ways of being we know; they delight and distress us, they cause us no end of anguish and disappointment; and, of course, we are beings ourselves. Let us stick to beings. What use is being?

Is the situation concerning the question of being somewhat as Nietzsche suggests (in his early period) when he says of Parmenides:

> Once in his life Parmenides, probably when he was very old, had a moment of the purest, completely bloodless abstraction undisturbed by any reality; this moment—un-Greek as no other in the two centuries of the Tragic Age—whose offspring is the doctrine of being—became the boundary-stone for his own life. (*Philosophy in the Tragic Age of the Greeks*, 1873)[2]

---

1. Compare summer semester 1932.
2. *Collected Works*, Musarion edition (ed. R. M. Oehler and F. Würzbach), vol. IV (Munich, 1921), p. 189.

And was the same Nietzsche correct when in his final period he said that being is "the last vapor of evaporating reality" (*Twilight of the Idols:* "Reason" *in Philosophy* § 4)?[3]

"Being": a thought un-Greek as no other—or indeed Greek?! "Being": vapor and smoke—or is it the innermost hidden fire of human Dasein? We do not know; for that reason we are questioning, that is, we are struggling to inquire correctly. All we know at the moment is that when beings are questioned at the very beginning, this differentiation of beings and being exists in the form of an equation. Here we must again note that τὸ ὄν actually signifies being [*das Seiend*] (participial) and being [*das Sein*] (beingness). We are provisionally and in a general way trying to throw light upon this peculiar state of affairs regarding the differentiation between beings and being so that we can at least surmise that we are not dealing here with a meaningless and arbitrary choice of words; all the more so because language is the source and wonder of our Dasein, and we may assume that philosophy did not misspeak at the time of its inception or when human beings came into their proper existence.

Τὸ ὄν = τὸ εἶναι; beings = being. We do not call any being at all, for example, this thing and that thing, by the name being; we do not even say *the* being, but *a* being. We call it this because this is the way we grasp and experience it. We experience this or that being without further considering that and how it is a being and for this reason belongs among beings. This gets taken for granted, so much so that we are not even aware of it.

But what do we mean by *the* beings, understood purely and simply? This and that being, the many things that are—plants, animals, human beings, human works, gods, all beings together, the complete itemized sum of individual beings—do philosophers mean this sum of individual beings when they say: beings—τὸ ὄν, τὸ εἶναι? Are then *the* beings the sum of all beings that we reach or try to reach by counting off and adding up the individual beings? Let us check. We begin to count—this being and added to it this being and then that and others, etc. And let us just assume we have come to the end. With what did

---

3. Op. cit., vol. XVII (Munich, 1926), p. 71.

§ 3. Equation or differentiation of beings

we begin? With an arbitrary being. Thus we did not begin with nothing so that we could fill this out by enumerating beings and thereby obtaining *the* beings. We begin with *a* being, and so indeed with a *being*. How is this? At the outset, until we have gone through and counted, we do not yet have *the* beings, assuming that *the* beings are formed by the sum. However, we begin in this way. We begin according to plan; in order to obtain the sum through counting, we begin with the sum. We select a first from the sum, which, according to our hypothesis, is in fact *the* beings; of course, the sum from which it is selected is taken as not yet counted up and tallied. Beings have their sum, the number of which, however, is not yet known; it is supposed to be determined in the count. We proceed from the uncounted sum and start off with an individual that belongs to this sum. So it is the uncounted sum from which we proceed to count; the sum is first. But that with which we start off counting and count as number one is not at all the first. In counting beings, we proceed from the uncounted sum. What is the uncounted sum? The so and so many, where the amount of the principal number still remains undetermined.

But do we really proceed from a numerically indeterminate sum when we count beings? Do we mean the beings, from which we proceed and out of which we select an individual, in the aforementioned character of numerical indeterminacy? Do we encounter the beings—out of which we seize and select countable individuals—as numerically indeterminate? Do we mean anything like this, and do we take the beings to signify the numerically indeterminate sum when we, for example, say that the sciences study beings and are divided into separate regions of beings? Obviously not. Beings are in no way numerically indeterminate. On the contrary, we encounter them as not at all numerical and thus also not indeterminate compositions. The beings in no way mean for us a kind of sum, be it determinate or indeterminate.

And yet—: the beings are for us something like the collective whole. What do we mean by the wholeness of this aggregate? We do not encounter it with regard to its determinate or indeterminate sum—although certainly "summarily." And we do take it this way and have taken it this way all along. Summarily, that is, on the whole, generally;

taken as whatever the single being is before any count, beyond every particular or universal. When we—and this continuously occurs as long as we exist—have beings before us, around us, in us, over us; when we become aware of these beings, then we are seized by that which intrudes and obtrudes itself on all beings from every side. Indeed, in the end, is it precisely this inherent *obtrusiveness* that causes us to address beings as beings, τὸ ὄν, and to say "it is," and to mean thereby quite explicitly what we call being, τὸ εἶναι? The wholeness of the whole (beings) is the original concentration of this obtrusiveness.

Individual beings do not first yield what we call the beings by means of summation; rather, beings are that from which we have always proceeded when counting off and adding up, whether or not we determine the number or leave it indeterminate. The beings permit the countability of individual beings; the sum of these, however, does not at all constitute being.

Beings, what are they? As what do they present themselves and present themselves to us? As that which we call being. First of all, above all and in every case beings are being: τὸ ὄν—τὸ εἶναι. Precisely when we take them as beings, beings are being. This is how we understand the equation: beings—being. This equation is already the first decisive answer to the question of what beings are, a response that required the most strenuous philosophical effort, in whose shadow all subsequent efforts pale. Thus at the same time we understand: When beings as such are asked about and when beings as such are made questionable, then being is questioned.

But what are beings? Now this means: What is being? The reply to this question is really just the complete answer to the question concerning beings. To be sure. And the first one we know of to have asked about beings in such a way as to have tried to comprehend being, and who also gave the first answer to the question, What is being? was *Parmenides*.

And what is he given to comprehend when he allows beings to obtrude in their obtrusiveness by questioning in this way? Precisely this *one* (the obtrusive present), such that he is not able to say anything else, but must say: τὸ ὄν τὸ ἕν—beings, they are precisely this one:

§ 3. Equation or differentiation of beings

being; being is the *one*, what beings as such *are*. He let himself be overwhelmed by this one, but greater still, he withstood this overwhelming power of the one and uttered each assertion about beings in terms of the gathered, simple clarity of this truth (see the later Antisthenes: the one and λόγος. Concerning τὸ ὄν and λόγος, see above, p. 5f.).

Parmenides bespoke the first decisive philosophical truth, and from that time onward philosophizing occurred in the West. The first truth—not only the first in time, the first to be found, but the first that precedes all others and shines throughout all that comes after. This is no "bloodless abstraction," no symptom of old age, but the gatheredness of thinking overladen with actuality. Nietzsche, who so surely ferrets out what underlies thinking and judging, never realized that his entire thinking was determined by this misunderstanding of Parmenides.

Ever since Parmenides, the battle over beings has raged, not as an arbitrary dispute about arbitrary opinions, but as a γιγαντομαχία, as Plato said, as a battle of titans for the beginning and end in the Dasein of the human. And nowadays—there remains but the wordplay of ambitious and clever chaps who purport to say that Parmenides' proposition—being is the one—is as false as it is primitive, that is, amateurish, awkward, and hence inadequate and of little worth. The falsity of a philosophical insight is, of course, an entirely different matter which we will not discuss further at this point. As to the primitiveness of the proposition τὸ ὄν τὸ ἕν, it is admittedly primitive, that is, original—in the strict sense of the word. In philosophy and indeed in every ultimately essential possibility of human Dasein, the beginning is the greatest, which henceforth can never again be attained; it is not only unable to be weakened and lessened by what follows but, if what follows is genuine, becomes truly great in its greatness and is expressly installed in its greatness. However, for those whose actions are oriented toward progress, the original and early become less and less important and real, and the most recent is *ipso facto* the best.

Even though Western philosophy up to Hegel has basically not gone beyond Parmenides' proposition: τὸ ὄν τὸ ἕν, despite all the transformations, this does not signify a deficiency but a superiority and indicates that in spite of everything, it remains strong enough to preserve its original truth.

(But it is preserved only as long as it still provokes and sustains a question. Nevertheless, this original truth is not that of essence; rather, essence is necessarily deprived of power in it. The beginning immediately becomes entangled in being as presence [actuality]; presence is the ineluctability [of essence] in the first breaking open.)

Thus by now it must be clear that the equation of τὸ ὄν and τὸ εἶναι is not an accidental, external, whimsical word choice but the first utterance of the fundamental question and answer of philosophy.

*

Beings, what are they? What is proper to them and only them? The answer is: being. Beings are meant here in the sense of beings as such. Ὄν ᾗ ὄν—in this ᾗ ὄν, beings are, so to speak, secured and retained only in order to show themselves and to say how things are with them. But we have not gotten very far with this discussion; in fact, quite the contrary. Being is here differentiated from beings. What are we differentiating? Distinguishing one being from another is fine. But being from beings?

What sort of peculiar differentiation is this? It is the *oldest difference;* there are none older. For when we differentiate beings from one another, that other differentiation has already occurred. Without it, even individual beings and their being different would remain hidden from us. A is differentiated from B—with the "is" we already maintain the older difference. It is the ever-older difference that we have no need to seek but find when we simply return (to remember: ἀνάμνησις). This oldest difference is, even more so, prior to all science and therefore cannot first be introduced through science and theoretical reflection about beings. It is merely espoused, cultivated, and used as self-evident by theoretical comprehension and in this way put into effect in everyday speech. This differentiation of beings and being is as old as language, that is, as old as human beings.

But in all this we are merely relating something about this differentiation, not comprehending it. Are we standing at the border of what is comprehensible? Is this difference ultimately the first concept? But then it must at least permit of demonstration and questioning precisely as to how the conceiving, that is, how the concept in its possibility, is determined by this boundary. Yet we ourselves know nothing of this. It has not yet even been questioned. On the contrary, we determine being the other way around—from the viewpoint of concept and assertion. For a long time the erroneous doctrine has existed that being means the same as "is," and that the "is" is said first of all in judgment. It therefore follows that we first understand being through judgment and assertion. While we are fond of appealing to the ancients in this connection (one of whose treatises it is our task to interpret here), I suggest that this errant opinion can appeal to the ancients only with partial legitimacy, which means with no legitimacy whatsoever.

§ 4. The manifoldness and unity of being

It is evident why Aristotle substitutes τὸ ὄν, beings, for εἶναι, being, about which he is inquiring: namely, because it stands for τὸ ὂν ᾗ ὄν, being. And being is one, ἕν. But does Aristotle not say that being is many and multifarious, πολλά, and thus πολλαχῶς? And is not this proposition the guiding principle of his entire philosophy? Is he also one of those who in the end no longer understand the insight of Parmenides? It would appear so, indeed it not only appears so but must obviously be so if we consider that Aristotle explicitly and in no uncertain terms battles against Parmenides. Aristotle emphasizes at *Physics* A 3, 186a22ff.: πρὸς Παρμενίδην, in relation to and against Parmenides is to be said ἡ λύσις . . . ψευδής, his solution (to the question of ὄν, that is, that ὄν, εἶναι is ἕν) is deceptive, it conceals the true; that is, ᾗ ἁπλῶς λαμβάνει τὸ ὂν λέγεσθαι, λεγομένου πολλαχῶς—Parmenides fails to understand the essence of being in that he assumes that beings are addressed ἁπλῶς, purely and simply as the simple one, whereas they must be understood in manifold ways. As proof of the manifoldness of beings (being), Aristotle mentions the πολλαχῶς of the categories (A 2, 185a21ff.), thus the πολλαχῶς in the narrow sense. This is worth noting. Yet it is not as though the *Physics* does not already have the πολλαχῶς in the wider sense; see Γ 1, 200b26f., where the discussion of κίνησις is most intimately connected with Book A.

Judging by what has been said, does Aristotle deny and disavow the first decisive truth of philosophy as expressed by Parmenides? No. He does not renounce it, but rather first truly comprehends it. He assists this truth in becoming a truly philosophical truth, that is, an actual question. Indeed, the πολλαχῶς does not simply push the ἕν away from itself; rather, it compels the one to make itself felt in the manifold as worthy of question. Those who believe that Aristotle merely added other meanings onto a meaning of being are clinging to appearances. It is a matter not just of embellishment but of a *transformation* of the entire question: the question about ὄν as ἕν comes into sharp focus here for the first time. Of course, it required first a decisive step over and against Parmenides. Plato undertook this, although admittedly at a time when the young Aristotle was already philosophizing with him, and this always means against him. Plato attained the insight that non-being, the false, the evil, the transitory—hence unbeing—also is. But the sense of being thereby had to shift because now the notness itself had to be included in the essence of being. If, however, ever since ancient times being is one (ἕν), then this intrusion of notness into the unity signifies its folding out into multiplicity. Thereby, however, the many (manifold) is no longer simply shut out from the one, the simple; rather, both are recognized as belonging together.

We modern chaps, with our short-lived but all the more clamorous discoveries, are hardly able any longer to assess the force of the philosophical effort that had to be exerted to discern that being as one is in itself many. But based on Plato's insight, it was once again an equally decisive step for Aristotle to discern that this manifoldness of being was multistructured, and that this structure had its own necessity. Aristotle's pointed confrontation with Plato stemmed directly from this. Whether Aristotle's πολλαχῶς represents only a continuation of Plato's later teachings that the one is many (ἕν—

§ 4. Manifoldness and unity of being 23

πολλά), or whether, conversely, the Platonic ἕν—πολλά represents the still-viable Platonic form of coming to grips with the already awakened Aristotelian πολλαχῶς by the elder Plato, will probably never be decided.

Since the basic relationship between Plato and Aristotle is undecidable, every popular pseudophilology that believes that for every thought of Aristotle's a prototype, no matter how sketchy or farfetched, can be found in Plato must be rejected. When someone asks the inane question, From whom did he get what he says here? believing himself thereby to be investigating the philosophy of philosophers, that person has already cut himself off from the possibility of ever being affected by a philosophy. Every genuine philosopher stands anew and alone in the midst of the same few questions, and in such a way that neither god nor devil can help if he has not begun to buckle down to the work of questioning. Only when this has happened can he learn from others like himself and thus truly learn, in a way that the most zealous apprentices and transcribers never can.

Yet, did we not assert, during the first enumeration of the four meanings of being in the Aristotelian sense, that the unity of these four meanings remains obscure in Aristotle? We did. However, this does not rule out but, for a philosopher of Aristotle's stature, precisely entails that this unity be troubling in view of its multiplicity. We need only observe how Aristotle explains the πολλαχῶς.

Thus he says on one occasion (*Met.* K [XI] 3, 1060b32f.): τὸ δ' ὂν πολλαχῶς καὶ οὐ καθ' ἕνα λέγεται τρόπον. "Beings are manifold and so not articulated according to one way." But he also sees immediately and clearly the result that this view, when taken out of context, could generate, namely the dispersion of ὄν into many τρόποι, a dissolution of the ἕν. In contrast, Aristotle states: παντὸς τοῦ ὄντος πρὸς ἕν τι καὶ κοινὸν ἡ ἀναγωγὴ γίγνεται (1061a10f.). "For each being, for all beings in whatever sense, there is a leading up and back to a certain one and common"; and at 1060b35: κατά τι κοινόν: "to some sort of common." We are always encountering this cautious and (as to what the encompassing one may be) openended τι (of some sort). Aristotle speaks of the final and highest unity of being in this fashion; see 1003a27 in *Met.* Γ (IV) 1, 1003a27 (and many other passages): τὸ ὄν ᾗ ὄν, τὸ εἶναι as φύσις τις—a sort of governing from out of and in itself.

Accordingly, Aristotle explicitly states: Being is said with an eye to something that is somehow common to all the various ways, and which cultivates a community with these so that these many are all of the same root and origin. *The ὄν is so little deprived of unity through the πολλαχῶς that, to the contrary, it could absolutely never be what it is without the ἕν.* Indeed, ὄν and ἕν are different conceptually, but in their essence they are the same, that is, they belong together. Aristotle gives precise definition to this original belonging together of the two. For example, in Γ 2 at 1003b22f.: τὸ ὂν καὶ τὸ ἕν ταὐτὸ καὶ μία φύσις τῷ ἀκολουθεῖν ἀλλήλοις. "Being and the one are the same and a (single) φύσις (a governing); for they follow one another." What Aristotle means by this is that each comes in the wake of the other; wherever the one appears, the other is also on the scene (see the παρακολουθεῖν in I 2 at 1054a14). Likewise in K 3 at 1061a15ff.: διαφέρει δ' οὐδὲν τὴν τοῦ ὄντος ἀναγωγὴν πρὸς τὸ ὂν ἢ πρὸς τὸ ἕν γίγνεσθαι. καὶ γὰρ εἰ μὴ ταὐτὸν ἄλλο δ' ἐστίν, ἀντιστρέφει γε· τό τε γὰρ ἕν καὶ ὂν πως, τό τε ὂν ἕν. "It makes no difference at all whether beings are traced back to being or to the one; for even though the two are not the same (that is, conceptually) but different, they still (that is, in this differing) face each other since the one is in some way or other being and being is one." Oneness belongs to the essence of being in general, and being is always already implied in oneness. Aristotle comprehends the peculiar character of this relationship as ἀκολουθεῖν ἀλλήλοις (ἀκολούθησις), as reciprocal following one another, and as ἀντιστρέφειν (ἀντιστροφή), as turning toward one another; ὄν and ἕν, so to speak, never lose sight of each other.

It may be difficult, even almost impossible, to truly shed light on this relationship through the texts of Aristotle that have been handed down to us. But just the same it remains indisputable that Aristotle lets the two of them be rooted in each other. The oneness of being is therefore rescued not only over against its manifoldness but precisely for it; rescued in the sense of the word as it was understood by Aristotle and Plato, to let something stand out as what it is, to not let it slip away and be covered over by the babble of common opinion for which everything is equally beyond question.

§ 4. Manifoldness and unity of being

Since τὸ ὄν and τὸ ἕν belong together in this way, it follows that λέγεται δ' ἰσαχῶς τὸ ὂν καὶ τὸ ἕν (I [X] 2, 1053b25). "Being and one are said in equally multiple ways."

It is certainly true, one might say, that Aristotle maintained the primordial affinity of being and oneness; certainly, one may further acknowledge, Aristotle also constantly refers at the same time to the πολλαχῶς. But nothing is thereby accomplished toward resolving the decisive question: How then is ὄν (εἶναι) ᾗ πολλαχῶς λεγόμενον, being as said in many ways, κοινόν τι, somehow held in common for the many?

Is this one being [*Sein*] something before all unfolding, that is, something that exists for itself, whose independence is the true essence of being? Or is being in its essence never not unfolded so that the manifold and its foldings constitute precisely the peculiar oneness of that which is intrinsically gathered up? Is being imparted to the individual modes in such a way that by this imparting it in fact parts itself out, although in this parting out it is not partitioned in such a way that, as divided, it falls apart and loses its authentic essence, its unity? Might the unity of being lie precisely in this imparting parting out? And if so, how would and could something like that happen? What holds sway in this event? (These are questions concerning *Being and Time*!)

Neither Aristotle nor those before or after him asked these questions, nor did they even seek a foundation for these questions as questions. Instead, only the various concepts of being and "categories" would later be systematized in accordance with the mathematical idea of science; see Hegel's comment in the second foreword to the *Science of Logic:* the material—is ready.

And yet Aristotle was also clearly concerned with the question of the unity of the πολλαχῶς λεγόμενα. For we find him attempting to respond to the question. And this attempt pressed against the very limit of what was at all possible on the basis of the ancient approach to the question of being.

We will see and understand this, however, only when we have first freed ourselves from the picture of Aristotle's philosophy drawn by the post-Aristotelian period up to and including our own age. The most catastrophic misinterpretation, fostered especially by medieval theology, was the following: The extremely cautious and provisional attempts at inquiring in the context of the truly guiding question concerning being were converted into primary, self-evident answers and main propositions of what was supposed to be Aristotle's philosophy. The question of being and unity became an axiom to be discussed no further: *ens et unum convertunter*—whatever is *ens* is *unum* and vice versa. Aristotle's treatises were thereby turned into a storehouse, or better yet, into a tomb containing such atrophied propositions.

The consequences of this complete covering over of the inner source which forms the basis of Aristotle's philosophy and ancient philosophy in general are still evident in Kant, even though it is he who tries to retrieve a genuine—though not original—meaning for the aforementioned teaching of scholastic philosophy. See *The Critique of Pure Reason* B113f.:

> In the transcendental philosophy of the ancients there is included yet another chapter containing pure concepts of the understanding which, though not enumerated among the categories, must on their view be ranked as *a priori* concepts of objects. This, however, would amount to an increase in the number of the categories and is therefore not feasible. They are propounded in the proposition, so famous among the Schoolmen: *quodlibet ens est unum, verum, bonum*. Now, although the application of this principle has proved very meager in consequences, and has indeed yielded only propositions that are tautological, and therefore in recent times has retained its place in metaphysics almost by courtesy only, yet, on the other hand, it represents a view which, however empty it may seem to be, has maintained itself over this very long period. It therefore deserves to be investigated in respect of its origin, and we are justified in conjecturing that it has its ground in some rule of the understanding which, as often happens, has only been wrongly interpreted. These supposedly transcendental predicates of things are, in fact, nothing but logical requirements and criteria of all knowledge of things in general, and prescribe for such knowledge the categories of quantity, namely unity, plurality, and totality. But these categories, which, properly regarded, must be taken as material, belonging to the possibility of the

things themselves, have in this further application been used only in their formal meaning, as being of the nature of logical requisites of all knowledge, and yet at the same time have been incautiously converted from being criteria of thought to being properties of things in themselves.

Kant knows only one alternative: to trace these determinations and relationships back to formal logic. However, if Kant is not understood in the way the Kantians understand him, and if one bears in mind that for Kant the original unity of transcendental apperception was the pinnacle of logic, and if this unity is not left simply hanging in the air but is questioned as to its own roots, then it can indeed be shown that and how Kant for the first time since Aristotle was once again starting to broach the real question about being.[1]

*

§ 5. Oneness of being—not as genus but as analogy

All of our earlier reflections have helped to map out and secure the realm of inquiry of Aristotle's treatise on δύναμις and ἐνέργεια. So far we have reached the following conclusions: (1) The question of δύναμις and ἐνέργεια is a question about ὄν (beings). (2) This inquiry concerning beings is fundamentally an inquiry concerning being (εἶναι). (3) Hence the inquiry about beings can be more precisely determined as a question about beings as such (ὄν ᾗ ὄν). (4) Being is the primary and decisive one that has to be said of beings; it is precisely, then, the reason that it itself is the one (ἕν). (5) But at the same time, being is said in various ways (πολλαχῶς). (6) Πολλαχῶς for its part is equivocal: (a) fourfold; (b) tenfold with respect to one of the manifold (the categories).

Now the final question of our preliminary considerations arises: How does Aristotle comprehend the unity of being as a manifold? Which πολλαχῶς, what sort of manifoldness, prevails in the ἀναγωγὴ πρὸς τὸ ἕν, in deciding the question concerning the oneness of the multiplicity? How is being understood in all this? If we can provide

1. See *Kant und das Problem der Metaphysik* (Bonn, 1929).

an answer to these questions, then we will have attained the gathered inner perspective for gauging the realm of the question—of the πολλαχῶς λεγόμενον.

If beings are addressed in various ways, then being is articulated in as many ways. Therefore, the word "being" has a multiple meaning: for example, being as being true or being possible or being at hand or being accidental; in each of these meanings, being is referred to. Or is it only the linguistic expression that the above-mentioned words have in common, whereas their meanings have absolutely no connection to each other? Just as [in German] the word *Strauss* can mean a bird, a bouquet of flowers, or a dispute. Here only the aural and written forms are the same in each, while the meanings and the things referred to are entirely different. Only the words are the same—a ὁμοιότης τοῦ ὀνόματος. Does the word "being" have only this sort of mere sameness of name? Are its many meanings (as Aristotle says) only said ὁμωνύμως? Clearly not; we understand being in being true and being possible in such a way that it expresses a certain sameness in each differentiation, even though we may be unable to grasp it. The being [*Sein*] that is expressed in the various meanings is no mere homonym; it implies a certain pervasive oneness of the understood significations. And this one meaning is related to the individual ways of being as κοινόν τι, as common. What then is the character of this κοινόν? Is the κοινότης, the commonness of the ὄν, somewhat like when we say, for example, "The ox is a living being" and "The farmer is a living being"? To both we ascribe what belongs to living beings in general. "Living being" is said of both of them, not simply ὁμωνύμως (*aequivoce*) but συνωνύμως (*univoce*). We do not here have simply a sameness of names; the two have the name in common, jointly, because each concretely defines what the name means through this one thing (living being). The farmer is a rational and the ox a nonrational living being; in contrast, *Strauss*, meaning a bouquet of flowers, is not a kind of bird.

What, then, is the status of the word "being" if it is not a mere ὁμωνύμον? Is being (ὄν) such a συνωνύμως λεγόμενον? That is, is being the unity of a highest genus (γένος) which we can get back to by separating out what is common from the various ways of being?

§ 5. Oneness of being

Aristotle says no, this too is impossible. The κοινότης of the ὄν is not that of a γένος. Why not?

In *Met*. B (III) 3, Aristotle offers a proof for this impossibility, that is, for the non-genus-like unity of being, and thus for the non-species-like character of the various ways of being and of the beings referred to in each of these ways. It is an indirect proof; why this is so, and indeed must be so, has to be shown later. The proof goes as follows: It is shown what would ensue were being to have the oneness and universality of a genus (as does living being); the result of this assumption proves to be an impossibility, and therefore the assumption from which it proceeded is also impossible; that is, the oneness and universality of being cannot be a genus.

Here is the proof in detail: Let us suppose that being is the genus for the different ways of being, and therefore for the individual beings that are each determined in their being by such ways. What is a genus? That which is universal and common to the many and can be differentiated and organized into species by the addition of specific differences. Genus is inherently related to species and thus to species-constituting differentiation; there is no genus in itself. For example, a genus is intended in the concept of living being. Plant, animal, human are species of this genus. It can be said of these in the same way, that is, in general, that they are living beings. That which characterizes a living being in general does not yet constitute what defines a human being as human, an animal as animal, a plant as plant, etc. In no way can the genus include things of this sort. Were the species-forming differentia already to be contained in it, it would simply not be a genus. For example, were the species-forming differentia which makes the human being a species of living being—rationality—to be included in the genus living being, then we could not declare plants and animals in general to be living beings; were we to attempt this, then it would have to be equally correct to say that plants are human. The content of the genus as such is necessarily independent of and uninvolved with the content of the species-forming differentia.

Let us suppose being to have the character of a genus and the different ways of being to be the species, for example, being true and being possible. Then the true and the possible would be of the sort that would have to be added to the genus "being" to form the species. "True" and "possible" would therefore have to add something to the genus "being" that, up to that point, it itself is not. Yet surely the true and the possible are not nothing at all but something; otherwise the genus "being" would be unable to determine them. But if they are something, τι, then they are the kind of things to which being belongs. Hence, the species-forming differentiae are beings of some sort in the broadest sense, that is, something that is determined by being. Therefore, being, which is attributable to anything that is not nothing, and indeed even to this, already expresses the species-forming differentiae, the true and the possible, as something that is. But for a genus to be able to be a genus, it may include nothing of the content of the species-forming differentiae. Since being, as what is able to be said most universally, must include this, it cannot have the character of a unity for the many in the manner of a genus; and the various ways of being cannot be understood as species.

Being cannot be a genus, cannot be said συνωνύμως. We can further extend our reflections on this proposition of Aristotle's. The universal, comprehended and defined as species-enabling genus, is usually called "concept." If being is not a genus, then it cannot be comprehended as a concept, nor can it be conceptualized. This is so not just because there is no higher genus than the genus of being, but also because being is not a genus at all. If the delimitation of a concept (ὁρισμός) is called definition, then this means that all definitional determinations of being must on principle fail. If being is to be comprehended at all, then it must be in a completely different way. We will find this explicated in the treatise on δύναμις and ἐνέργεια.

The proof of the non-genus-like character of being tells us only which characteristics the oneness of being is not. Had Aristotle done nothing but work out this negative answer, it would be proof enough that the question of the oneness of being was a real question for him. Since for both the earlier and the later Plato all determinations of being and being itself remained γένος, the Aristotelian statement of the problem announces a fundamental rejection of the Platonic conception of being—and a decisive step closer to the essence of being. (Whether the cited indirect proof is itself intrinsically valid will not be discussed here.)

§ 5. Oneness of being

Being is said neither ὁμωνύμως nor συνωνύμως. And yet in each instance it is understood and said as κοινόν τι; in fact, Aristotle even says (I 2, 1053b20f.): τὸ ὂν καὶ τὸ ἓν καθόλου κατηγορεῖται μάλιστα πάντων. "Being, like the one that goes together with it, is what is for the most part said of the whole (for the most part in general of all)." It is the foremost and ultimate κατηγόρημα; indeed, as Aristotle says in this chapter, it is κατηγόρημα μόνον, the most general declaration [*Gesagtheit*] and this alone. Nevertheless, it is to be noted that Aristotle never characterizes being as a category, albeit that being holds sway in the categories; ἡ κατηγορία, in the sense we have already encountered (Δ 7; see above, p. 11).

How then are we to understand the relationship of this most general one to its many different ways? Is there in fact any relationship of the general to the many encompassed by it which is other than the relation of genus to species and the particulars it unites? There is such a relation. But Aristotle nowhere shows it directly in the relationship of being to the multiplicity of ways of being; instead he turns his attention to a peculiar kind of meanings in language which express a oneness of many without being a genus for this unified many (Γ 2, beginning).

For example, the word "health"; it is the general designation for the healthy as such. We say that someone has a healthy heart; "healthy" in this case expresses the characteristic of a specific condition of the body. The body is healthy because it has incorporated the condition that is named and because it can in general possess this condition (δεκτικόν). We also say that a medicinal herb is healthy, but we do not mean by this that the condition of the plant in question is not sick; rather, the medicinal herb is healthy because in some cases it produces health (ποιεῖν). Then again, we also say that a person has a healthy complexion, and we do not mean that the complexion is healthy and not sick—a hue can be neither healthy nor sick; rather, what is meant is that the complexion is an indication (σημεῖον εἶναι) of health in the first sense, understood as a physical condition. Furthermore, we say that a walk is very healthy. Here again, we do not mean that walking is the very opposite of sick; nor do we mean that walking is a sign of health; nor do we mean that it produces health. Rather, it is healthy because it is conducive to the recovery and improvement of health (φυλάττειν). Thus "healthy" is said of the heart, of medicinal herbs, of the complexion, of walking; all four are healthy, and yet they cannot be called this in the same sense. Health (healthiness) is predicated of many different things. And yet it is not the genus for the many; otherwise health would have to be predicated of the many in the same general way, whereas it is precisely (in contrast to γένος, it is a ἑτέρως λεγόμενον) predicated in different ways of heart, herb, complexion, and walking, so much so that the meanings of healthiness mentioned in the second, third, and fourth places are each related differently to healthiness in the first sense. They necessarily co-signify the first sense: medicinal herbs as bringing forth health, the healthy complexion as a sign of health, taking a walk as maintaining health.

We can infer from this first of all that already in language itself there are peculiar relationships that apparently are expressed in a logical form. But we surmise from the kind of meaning that here ordinary logic surely breaks down. Language itself can in no way be understood logically—a fact that we are only now gradually realizing. We must free the categories of **language** from the framework of logic that has ruled over it since the **time** of the Alexandrians, prefigured, of course, in Plato and Aristotle. We can clarify the extent to which the relationships of meaning develop among themselves in manifold ways through yet another meaning of "healthy." When we say that a sound thrashing is sometimes very healthy, we are not conveying a fifth meaning of "healthy" that is structurally similar to the other four previously mentioned. True, a sound thrashing does refer to the body; however, not in the sense that it fosters health. In fact, quite to the contrary. "Healthy" is here meant as beneficial to one's formation, that is, precisely not so much having to do with the body. "Healthy" in this sense is a metaphorical use of the meaning according to which, for example, we call taking a walk "healthy."

Despite the diversity of these different meanings, they do have a unity. What is the character of this unity?

Aristotle at one point briefly discusses difference and the unity of what differs that pertains to it—on the occasion of the delineation of the essence of sameness (ταὐτόν, Δ 9). He states there (1018a12f.):

§ 5. Oneness of being

διάφορα λέγεται ὅσ' ἕτερά ἐστι τὸ αὐτό τι ὄντα, μὴ μόνον ἀριθμῷ, ἀλλ' ἢ εἴδει ἢ γένει ἢ ἀναλογίᾳ. "Different are all things that differ (among themselves) while nevertheless in some way remaining the same, not only numerically but with regard to species or genus or analogy." In addition to the numerical unity of many different things and the unity of species and genus, Aristotle recognizes the *unity of analogy*. What is meant by this?

The signification "healthy" contains a unity for different things, namely, the kind whereby the primary meaning—"healthy" as a characteristic of one's physical condition—takes on the function of unifying the other meanings in that it lets these other meanings be related to itself, each in a different way. These different meanings are in keeping with and comply with the first, each in a specific regard. However, the primary meaning is not the genus of the others; there is absolutely no universal meaning of "healthy" that could be suspended over the various meanings that have been mentioned and yet still say something. Just as a person's complexion is healthy by virtue of its being an indication of healthiness in the primary sense, so correspondingly—but not in the same way—taking a walk is healthy with respect to the promotion of health in the primary sense; the latter appears and is sustained in various relationships, and what belongs to this somehow corresponds to being healthy. "Healthy" does not directly express something about the physical condition, but the meaning corresponds [*ent-spricht*] to it, takes it into consideration, has regard for it, just as we sometimes say: a request has been taken into consideration, it is in correspondence, it is accepted.

This corresponding, ἀναλέγειν, is intrinsically an ἀναφέρειν πρὸς τὸ πρῶτον (compare Γ 2, 1004a25): a "carrying onto the first" of the meaning and securing it there. This πρῶτον is ἐξ οὗ τὰ ἄλλα ἤρτηται, καὶ δι' ὃ λέγονται (1003b17), "that upon which the other meanings are hinged and secured and that through which the other meanings can be (understood and) said." The manner of the carrying back and forth of the meanings to the first is different in each case. The first, however, is the *sustaining and guiding basic meaning;* it is always that from out of which the meaning which carries itself to it and corresponds [*ent-sprechen*] to it is capable of being spoken. In Greek, the "from out of which" is the ἀρχή; it is for this reason that Aristotle generally defines the essence of ἀναλογία as λέγειν πρὸς μίαν ἀρχήν (see 1003b5f.). This ἀρχή is that which unifies the many that correspond to it, that is, the sustaining and guiding meaning to which the various ways of specific corresponding in each case correspond. The λέγειν of the λόγος of the ἀναλογία is the λέγειν πρὸς ἕν—πρῶτον. This ἓν πρὸς ὅ is therefore a κοινόν; not the simple κοινόν of the γένος but κοινόν τι—some sort of common, that which is inherently there, as a mode of the same, to hold together the corresponding in a unity.

This is a preliminary explanation of analogy as a mode of unity. We will have an opportunity at a later point to delve more specifically into the matter of the essence of analogy. Characteristically, Aristotle does not clarify here (Γ 2) the analogical character of being (ὄν) in a direct manner, but once again through an analogy.

*

§ 6. The questionableness of the analogy of being

We wish today to conclude our preliminary considerations. It was a matter of delimiting in a general way the realm of questioning in which our treatise belongs, that is, being and the manifoldness of its modes. We asked in conclusion: How did Aristotle conceive the unity of being as a manifold, and which πολλαχῶς was the leading one in the ἀναγογή πρὸς τὸ ἕν? It was necessary to show how in general a meaning is *one* with regard to the many that belongs to it: whether ὁμωνύμως or συνωνύμως, or even (although Aristotle does not use this expression) ἀναλογικῶς. For ὄν is said neither ὁμωνύμως nor συνωνύμως (as the γένος), and yet it is said as κοινόν, in general, even καθόλου μάλιστα πάντων. How, then, is the unity of this universality of being to be conceived as a sort of analogy? The unity of the meanings of "health" is an example of analogy. "Being" signifies in a way that corresponds to the way "health" signifies.

Now it must be shown how Aristotle establishes the unity of anal-

§ 6. Questionableness of analogy of being

ogy as that unity according to which ὄν, ἕν, and κοινόν τι belong to the πολλαχῶς λεγόμενα. The λέγεσθαι of this πολλαχῶς is the λέγειν of ἀναλογία. Accordingly, the question arises: πρὸς τί λέγεται τὰ πολλαχῶς λεγόμενα—with respect to what? It must be a πρῶτον and an ἀρχή, and, since what is at issue is ὄν, it must be the πρῶτον ὄν or the ὄν πρώτως λεγόμενον. Thus what is being sought is the sustaining and leading fundamental meaning of ὄν, of being, πρὸς ὅ τὰ ἄλλα λέγεται—with respect to which the others are said. What is this?

We have evidently already gotten to know it. We need now only read with a more refined understanding the first sentences from Θ 1, from which our entire introductory considerations have arisen. Περὶ μὲν οὖν τοῦ πρώτως ὄντος καὶ πρὸς ὃ πᾶσαι αἱ ἄλλαι κατηγορίαι τοῦ ὄντος ἀναφέρονται εἴρηται, περὶ τῆς οὐσίας. We translated it at that time conservatively: "[We have dealt with] beings in the primary sense. . . ." Now we can translate it in the following way: "We have dealt with the sustaining and leading fundamental meaning of being, to which all the other categories are carried back (ἀναφέρονται, we could equally well say: ἀναλέγονται, are said back), that is, οὐσία." The first category is the sustaining and guiding fundamental meaning of being and as such the κοινόν, which imparts itself to all the others so that these themselves have the meaning of being due to their relationship to it. But it is well to note that οὐσία as this ἕν and πρῶτον is not κοινόν in the sense of a genus which is named and said of the other categories as species. Being so constituted and being so much are not kinds of οὐσία but ways of being related to it.

Being so constituted, for example, signifies a way of being; and since the being so constituted is the being so-constituted of something, this being is related to οὐσία. The being such is not, however, a kind of οὐσία being, but πρὸς τοῦτο, so much so that the οὐσία is always said along with it (just as along with the various meanings of "healthy" the first meaning is included).

We have already mentioned that Aristotle over and against Plato secured another ground with his question of the unity of being and had to critically reject the doctrine of the ideas (as γένη). (Insofar as the expressions γένος and εἶδος play a role in Aristotle, they have a transformed meaning.) The import of this position is shown, for example, in his handling of the question of the idea of the good, which for Plato was decisive. Aristotle says: There is no idea of the good, or better, the good does not exist as idea and highest γένος, it does not have the character of an idea; for τἀγαθὸν ἰσαχῶς λέγεται τῷ ὄντι, "the good is understood exactly as variously as being" (see *Nicomachean Ethics* A 4, 1096a23ff.); the individual categories, according to which what we call good is good and can be good, are now enumerated: thus, for example, good in the sense of τί (what-being) can be god or νοῦς; in the sense of being so constituted it can be ἀρετή; or in the sense of being at the time, καιρός, the right moment. And then he says: δῆλον ὡς οὐκ ἂν εἴη κοινόν τι καθόλου καὶ ἕν, "it is clear from this that there is no universal and one," that is, such a one that would hang over everything as the highest genus. Aristotle summarizes this problem in a form that makes clear that the entire question of oneness is oriented toward the analogy of being: οὐκ ἔστιν ἄρα τὸ ἀγαθὸν κοινόν τι κατὰ μίαν ἰδέαν. ἀλλὰ πῶς δὴ λέγεται· οὐ γὰρ ἔοικε τοῖς γε ἀπὸ τύχης ὁμωνύμοις (1096b25ff.). "Thus the good is not some sort of commonality (pursuant to) with regard to an idea. But then in what manner is it said? For it is not like that which only accidentally has the same name." This thought is important in that Aristotle here states: not merely not ὁμώνυμν but not ὁμώνυμν ἀπὸ τύχης, not a mere accidental homonym. This ἀπὸ τύχης (accidentally) occurs here because Aristotle does in fact sometimes say (see *Met.* Δ 12, 1019b8): τὸ ὂν ὁμωνύμως λέγεται—being is used in the sense of a homonym. This is meant first of all only negatively: not συνωνύμως, not as genus; and what is not συνωνύμως is a ὁμωνύμως. This, then, is here to be understood as something which nevertheless has meaning in some way or other, as a meaning which is certainly not συνωνύμως, yet has a real unity of meaning. Being is not purely and simply an accidental ὁμώνυμον, but a sort of one, in the sense of analogy. Hence the question: ἀλλ' ἀρά γε τῷ ἀφ' ἑνὸς εἶναι ἢ πρὸς ἓν ἅπαντα συντελεῖν, ἢ μᾶλλον κατ' ἀναλογίαν· ὡς γὰρ ἐν σώματι ὄψις, ἐν ψυχῇ νοῦς, καὶ ἄλλο δὴ ἐν ἄλλῳ (*Nicomachean Ethics*, A 4, 1096b27ff.). "Is it (this one) perhaps similar to being from one, or similar to the fact that all (meanings)

§ 6. Questionableness of analogy of being 37

converge into one (as their end), or is it rather similar according to 'analogy'? For just as the eyes are what sees in the body, so is νοῦς in the soul and so others in other cases." Here the κατ' ἀναλογίαν (according to analogy) is differentiated, by means of the comparing but excluding ἤ—ἤ, from the ἀφ' ἑνός (from one; compare ἐξ οὗ τὰ ἄλλα ἤρηται, p. 33 above) and from the πρὸς ἓν συντελεῖν (the converging into one). It should also be observed that in this same passage another concept of analogy is presented that is *not* identical to the categorial relationships (see p. 48f. below).

However little all this may be clear in the end, we see in what direction Aristotle positively seeks the oneness of the ὄν for the multiplicity. And thus the oneness of the realm of the questioning ought to be able to be determined, that is, the how according to which the ὄν πολλαχῶς λεγόμενον is—ἕν. However, we recall: Aristotle uses the πολλαχῶς in a broad and in a more restricted sense. What we have just now been discussing is the πολλαχῶς in the *more restricted* sense in which the multiplicity of the categories is meant. But all the categories together with the first still make up only one of the τετραχῶς within the πολλαχῶς in the broad sense.

Already in the Middle Ages, on the basis of the above sentence from the beginning of *Met*. Θ 1, it was concluded that the first guiding fundamental meaning of being in general—for the four ways together as well, not only for the one and its multiplicity—was οὐσία, which is usually translated as "substance." As if being possible and actual and true also had to be led back to being in the sense of substance. They were even more inclined to conclude this in the nineteenth century (especially Brentano), since in the meantime, being, being possible, and being actual had come to be perceived as categories. Hence it is a generally accepted opinion that the Aristotelian doctrine of being is a "substance doctrine." This is an error, in part resulting from the inadequate interpretation of the πολλαχῶς; more precisely: it was overlooked that only a question is here first of all being prepared. (W. Jaeger's reconstruction of Aristotle is built upon the basis of this fundamental error.)

And so now for the first time the decisive question arises: What is the kind of unity in which this *broad* πολλαχῶς is held together (that is, τὸ ὄν κατὰ τὰ σχήματα τῆς κατηγορίας, κατὰ δύναμιν ἢ ἐνέργειαν, ὡς ἀληθὲς ἢ ψεῦδος, κατὰ συμβεβηκός)? Is the unity here also that of analogy? And if so, what then is the πρῶτον ὄν, πρὸς ὃ τὰ τέτταρα λέγεται? What therein is the φύσις τις, that which is determined and governed from out of itself? Here everything becomes obscure. We always find only the itemizing juxtaposition. And in addition we find the claim: the ὄν has for its multiplicity the unity of analogy.

The analogy of being—this designation is not a solution to the being question, indeed not even an actual posing of the question, but the title for the most stringent aporia, the impasse in which ancient philosophy, and along with it all subsequent philosophy right up to today, is enmeshed.

In the Middle Ages, the *analogia entis*—which nowadays has sunk again to the level of a catchword—played a role, not as a question of being but as a welcomed means of formulating a religious conviction in philosophical terms. The God of Christian belief, although the creator and preserver of the world, is altogether different and separate from it; but he is being [*Seiende*] in the highest sense, the *summum ens;* creatures—infinitely different from him—are nevertheless also being (*seiend*), *ens finitum*. How can *ens infinitum* and *ens finitum* both be named *ens*, both be thought in the same concept, "being"? Does the *ens* hold good only *aequivoce* or *univoce*, or even *analogice*? They rescued themselves from this dilemma with the help of analogy, which is not a solution but a formula. Meister Eckhart—the only one who sought a solution—says: "God 'is' not at all, because 'being' is a finite predicate and absolutely cannot be said of God." (This was admittedly only a beginning which disappeared in Eckhart's later development, although it remained alive in his thinking in another respect.) The problem of analogy had been handed down to the theology of the Middle Ages via Plotinus, who discussed it—already from that angle—in the sixth *Ennead*.

The first and ultimate πρῶτον ὄν, πρὸς ὃ τὰ ἄλλα λέγεται, which is thus the first meaning for the πολλαχῶς in the broad sense, is obscure. And therefore the πρώτη φιλοσοφία, genuine philosophizing, is inherently questionable in a radical sense. All this is later erased by the thesis that being is the most self-evident. (This questionability is a far stretch from the image that is commonly held of Aristotle when his philosophy is envisioned along the lines of the scholarly activity of a medieval scholastic or German professor.)

So we also do not know how the ὄν as δύναμις and ἐνέργεια stands in relation to the other meanings or how it stands together with them in the unity of being. And it is precisely here that we must guard against manipulating things artificially in order to concoct in the end a smooth "system." It is necessary to leave everything open and questionable; only thus will we be capable of freeing and keeping alive Aristotle's unresolved innermost questioning, and thereby the questioning of ancient philosophy and accordingly our own. What are beings as such? What is being such that it unfolds itself four times? Is the fourfoldness at which Aristotle directs the being question the most original fourfold of being? If not, then why not? Why does Aristotle chance upon precisely this number of four? How is being understood in all ancient questioning such that it extends itself into the realm of questioning that we find here?

The treatise which we have made the object of our interpretation stands in the obscurity in which we grope with the posing of these questions. And we are taking up this treatise on the basis of an initially unfounded conviction that precisely this treatise, when we follow it philosophically, lets us advance the farthest into this obscurity—that is to say: it forces us into the basic question of philosophy, presuming we are strong enough to let ourselves be truly compelled.

Chapter One

Metaphysics Θ 1. The Unity of the Essence of Δύναμις κατὰ Κίνησιν, Force Understood as Movement

§ 7. Considerations for the movement of the entire treatise on δύναμις *and* ἐνέργεια

This treatise which concerns δύναμις and ἐνέργεια takes its bearings within the guiding question of philosophy: What are beings as such? It attempts for its part and in its own fashion to achieve a clarification of being. What course does it follow? Where does it begin? This is stated in the following sentences (1045b35-1046a4):

> καὶ πρῶτον περὶ δυνάμεως ἣ λέγεται μὲν μάλιστα κυρίως, οὐ μὴν χρησίμη γ'ἐστὶ πρὸς ὃ βουλόμεθα νῦν. ἐπὶ πλέον γάρ ἐστιν ἡ δύναμις καὶ ἡ ἐνέργεια τῶν μόνον λεγομένων κατὰ κίνησιν. ἀλλ'εἰπόντες περὶ ταύτης, ἐν τοῖς περὶ τῆς ἐνεργείας διορισμοῖς δηλώσομεν καὶ περὶ τῶν ἄλλων.

"And first (we want to treat) δύναμις in the sense in which the word is most properly used; admittedly δύναμις understood in this way is not truly needed for what we now have before us (in this treatise). For the δύναμις and the ἐνέργεια (that are properly our theme) extend further than the corresponding expressions which are taken only with regard to movement." Here we are implicitly to think along with this: we are first dealing only with δύναμις in its usual most readily available meaning. "But after we have dealt with this, we want to open up the others (namely, the more far-reaching meanings of δύναμις) in our discussions of ἐνέργεια."

These sentences are of decisive importance for understanding the point of departure and the inner workings of the question which the treatise as a whole poses. We first learn quite generally: δύναμις and ἐνέργεια are, on the one hand, μάλιστα κυρίως λεγόμενα—under-

§ 7. Treatise on δύναμις and ἐνέργεια

stood most prevalently—but then, on the other hand, ἐπὶ πλέον—extending further. First taken according to how they are commonly and predominantly understood, thus according to how they are first of all understood at any given time. But how exactly? Answer: κατὰ κίνησιν (ἡ κατὰ κίνησιν λεγομένη)—with regard to movement. How is this to be understood?

When we *regard movements*, we encounter what moves. And then we speak (without further ado) of forces which move what moves, and likewise of activities at work and in work (ἔργον). The Greek "ἔργον" has the same double sense in which we use the German *Arbeit* (work): (1) work as occupation, as when we say, for example, "He didn't make the most of his working time"; (2) work as what is diligently worked upon and gained through work, as when we say, "he does good work." Ἐνέργειαι are the activities, the ways of working (ἔργα in the first sense), which are occupied with work (ἔργον in the second sense): the ways of being-at-work. It is necessary to hear this double sense: precisely to be caught up in enactment and so to have something to produce. When we encounter what moves, we speak of forces and activities which are themselves related to movement, to the moving of what moves: κατὰ κίνησιν. In what follows, δύναμις κατὰ κίνησιν is to be defined. The κατά implies an inner reference to δύναμις itself. Κατὰ λεγομένη means then: being in this way and therefore addressable in this respect as well.

But now it must be noted: We speak of forces and activities in the plural (δυνάμεις, ἐνεργείαι); there are many kinds of such forces and activities which indeed correspond to the many kinds of beings that move and which, like these beings, are also present. But over and against these present forces and activities there is ἐπὶ πλέον: ἡ δύναμις καὶ ἡ ἐνέργεια—*the* δύναμις and *the* ἐνέργεια, in the singular, stated simply and understood singularly, uniquely. We translate ἐπὶ πλέον: the δύναμις and ἐνέργεια, taken singularly, extend "further." This means: over a broader realm. And yet this cannot then mean that outside the circle of what moves we would find still other forces and activities as well. Instead, the δύναμις and the ἐνέργεια in the singular mean a πλέον in the sense of something higher and more essential. Δύναμις and ἐνέργεια ἐπὶ πλέον do not therefore signify simply an extension of the realm of applicability but rather an essential transformation of the meaning and thereby, of course, a fundamental furthering of the reach of this meaning. This ἐπὶ πλέον is ὃ βουλόμεθα—that which we want above all to expose. And just this exposition of δύναμις and ἐνέργεια ἐπὶ πλέον is the decisive, basic discovery of the entirety of Aristotelian philosophy; δύναμις and ἐνέργεια, taken singularly, obtain for the first time through philosophical inquiry an essentially other, higher meaning.

This meaning arises from within a philosophical inquiry. But this inquiry takes place under the guiding question: τί τὸ ὂν ἧ ὄν—what are beings as such? The essential meaning of δύναμις and ἐνέργεια arises therefore, to state it now negatively, not κατὰ κίνησιν—not when we let what moves be encountered as present and notice as well what is also commonly present along with it—that is, not when the present being that moves is seen as referring to a moving present force, nor conversely, when it is seen from out of this force. And so Aristotle says at a later passage (chap. 3, 1047a32): δοκεῖ γὰρ [ἡ] ἐνέργεια μάλιστα ἡ κίνησις εἶναι—movement appears to be something like a being-at-work. The most obviously general character of κίνησις is ἐνέργεια. To what extent? Where something is in movement we do say: here something is underway, something is afoot, at work; here is an activity.

The essential meaning of δύναμις and ἐνέργεια, on the contrary, is not rendered κατὰ κίνησιν, or, stated more carefully, is not rendered κατὰ κίνησιν μόνον—not only with regard to movement. How is this to be understood? What is meant here becomes only more obscure if we consider that Aristotle achieves this essential meaning of δύναμις and ἐνέργεια precisely through a treatment of κίνησις, with a view toward movement. This is shown quite unmistakably in his investigation of κίνησις (*Phys.* Γ 1-3).

*

The horizon of questioning for this inquiry into δύναμις and ἐνέργεια is being and its unity in πολλαχῶς. The unity of being is set forth as the unity of analogy. The unity of horizon and the interpretation of being get

§ 7. Treatise on δύναμις and ἐνέργεια

lost in obscurity. Since: (1) The essence of analogy is not properly clarified. (2) The analogous character is not demonstrated but rather only illustrated by means of the analogy "healthy" and other things like this; it is not shown in what manner the fourfold divided being is to be unified in correspondence to one dominating and guiding meaning. (3) It is not shown which meaning among the four this is. (4) It is not shown why being is divided into a fourfold. (5) It is not shown why it must have the unity of analogy, nor why only the indirect proof of *Met*. B 3 can be given for this. (6) Neither is it shown that this horizon of the inherently analogous being is necessarily the most far-reaching and why this is the case. (7) The problem of a transcendental horizon is not here at all—the understanding of being as such is not seen; there is only the juxtaposition of λόγος and ὁρισμός, of κίνησις and παρουσία, of τέλος and so on.)

(To what extent, then, is the characteristic of being as πολλαχῶς λεγόμενον an essential one? Does this state positively that being is inherently multiple? Is it multiple in itself or for us? Or is it neither of these but rather more originally, in its essence, which would still pertain, of course, to both of the above? The πολλαχῶς is thus a symptom of disempowerment: essence undergoes corruption [*Wesen verwest*], and for this reason being consists in a "present" multiplicity; this multiplicity as such remains misunderstood and unquestioned.)

—Our interpretation commences with the question, How does Aristotle characterize the point of departure and the inner workings of his inquiry concerning δύναμις and ἐνέργεια? We now have the following distinction: δύναμις and ἐνέργεια in their most usual meaning, which have at the same time a plural form, and then in addition a further, that is a higher and more essential, meaning, which can be used only in the singular. From this emerges a double difficulty. The ordinary meaning is used κατὰ κίνησιν μόνον: the forces found along with present movements. This meaning is for the proper aim of the inquiry οὐ χρησίμη, not needed—and yet it is precisely these forces which are dealt with so extensively. The essential meaning of δύναμις and ἐνέργεια is οὐ κατὰ κίνησιν, not the current meaning understood with a view toward movement—and yet this meaning again becomes the issue precisely in the context of the inquiry into κίνησις, the phenomenon of movement.

Then is the singular meaning of δύναμις and ἐνέργεια nevertheless not to be taken κατὰ κίνησιν? Certainly not. For to question κατὰ κίνησιν and to take δύναμις as κατὰ κίνησιν is fundamentally different from questioning κατὰ κινήσεως (genitive), from asking whether δύναμις has anything at all to do with *movement as such*—not only to ask to what extent does any δύναμις whatsoever move what moves and bring about movement, but whether movement as such is determined by δύναμις. Let us take an example which Aristotle often uses: When a house is built, all sorts of things are in movement. Stones and beams are laid upon one another, coming together to form the work. Forces and activities are also at work here. If we look upon this movement as a whole and discern the activities and forces which are here present, we are then viewing κατὰ κίνησιν and so also perceiving δυνάμεις, those things which are also present along with what moves, namely, along with those present beings in movement. But we are not viewing here movement as movement, not viewing κινούμενον ᾗ κινούμενον; we are not asking what moved-being as such would be. We are not taking the κινούμενον as ᾗ ὄν, and we are not taking the κίνησις as ᾗ εἶναι. We are not dealing κατὰ κινήσεως, with movement, so that it as such is the theme. To question in this fashion would be to ask about εἶναι, being, and thereby about δύναμις and ἐνέργεια, but in a completely different sense (ἐπὶ πλέον). If, accordingly, in this treatise δύναμις and ἐνέργεια ἐπὶ πλέον are supposed to be the theme of the inquiry, this does not then exclude that κίνησις remain in view; on the contrary, it must remain in view but not κατὰ κίνησιν.

And yet the inquiry is supposed to treat δύναμις first κατὰ κίνησιν. And not only this, it is to treat δύναμις κατὰ κίνησιν despite this οὐ μὴν χρησίμη (Θ 1, 1045b36)—not truly needed for what the treatise properly intends. A remarkable method for an Aristotle whose rigor and acumen remain unrivaled by all subsequent investigative philosophical questioning, with the exception of Kant.

We now have the following state of affairs: δύναμις κατὰ κίνησιν is being dealt with despite its not being needed for the clarification of δύναμις κατὰ κινήσεως. The discussion concerning δύναμις κατὰ κίνησιν will indeed then have to be in some sense a preparation for

§ 7. *Treatise on* δύναμις *and* ἐνέργεια 45

the proper inquiry. What results in reference to this δύναμις is, however, not needed for the δύναμις in its essential meaning; it does not constitute a determinant element of this δύναμις. And yet the inquiry is nonetheless useful and helpful. One may therefore by no means interpret and translate οὐ χρησίμη as pointless or useless, but rather as not needed, inapplicable, not to be taken over into the essential concept. Precisely because the δύναμις κατὰ κίνησιν is inapplicable but still δύναμις, and on the other hand generally oriented toward κίνησις, precisely for this reason is the inquiry regarding it useful for what is here decisive. In stepping away from this δύναμις, the step toward δύναμις ἐπὶ πλέον is accomplished. From it, the leap to ἐπὶ πλέον can be achieved.

But if δύναμις κατὰ κίνησιν is in this way helpful, then why is the inquiry not designed so as to proceed from δύναμις in its usual sense to δύναμις in its essential meaning? Aristotle proceeds otherwise; he says, in fact, explicitly (1046a3): δύναμις in its essential meaning is first to be dealt with ἐν τοῖς περὶ τῆς ἐνεργείας διορισμοῖς—first in the discussions concerning ἐνέργεια. Thus he states at the beginning of chapter six (1048a25f.): ἐπεὶ δὲ περὶ τῆς κατὰ κίνησιν λεγομένης δυνάμεως εἴρηται, περὶ ἐνεργείας διορίσωμεν. "Now that δύναμις κατὰ κίνησιν (δύναμις in its usual meaning) has been dealt with, we want to take up the inquiry concerning ἐνέργεια." Does the advance from the first point of departure via δύναμις as it is ordinarily understood occur in such a way that from it a transition is made to the essential ἐνέργεια? That is not what is said here. The possibility remains that the movement in the subsequent sections proceeds as follows: originating from δύναμις κατὰ κίνησιν, advancing to ἐνέργεια κατὰ κίνησιν, passing over to the ἐνέργεια κατὰ κινήσεως, and proceeding to the δύναμις κατὰ κινήσεως.

This would be a characterization of the inner movement of the inquiry; presumably this is required by the matter itself. Whether this is the case is something we will have to decide only as the matter is brought closer to us. A further question remains as to whether this movement can be carried out in actuality with such a separation into individual sections. And if not, why not?

However the movement may be in actuality, at this point it is not easy to comprehend why the inquiry into δύναμις and ἐνέργεια takes as its point of departure δύναμις at all. Because it belongs to τὰ ἡμῖν γνώριμα (cf. Z 3, near the end)? Only for this reason? Enough—we shall now pursue the point of departure.

Still one more prefatory remark is needed. The δυνάμεις κατὰ κίνησιν are the present forces of which we speak when we are confronted by whatever moves. If these δυνάμεις are now to be dealt with, then this is so not in the sense that Aristotle simply attempts to establish which δυνάμεις actually occur; rather, it is to be asked what δυνάμεις as such are: to be asked περὶ δυνάμεως. This preliminary inquiry also asks about a being as such and is therefore philosophical. It is not as if the philosophical inquiry first begins only where ἡ δύναμις καὶ ἡ ἐνέργεια ἐπὶ πλέον (in their essential meaning) make their way into the discussion.

*

The relationship of δύναμις and ἐνέργεια κατὰ κίνησιν to ἡ δύναμις and ἡ ἐνέργεια which are ἐπὶ πλέον, this transition, is not simply the exchange of one for the other, but is rather originally one, a project with its foothold in δύναμις κατὰ κίνησιν and ἐνέργεια κατὰ κίνησιν. (This transition is thus neither a mere extension nor an abstract universalization, but the transformation of the question that proceeds from κίνησις to a question about something that is to be accounted for from out of itself, that is, on the basis of the essence of being as what is indissociable from it.) But then why does the inquiry go in Θ 6 from δύναμις κατὰ κίνησιν directly to ἐνέργεια ἐπὶ πλέον and to the corresponding δύναμις ἑτέρως? Which ἐνέργεια corresponds to the δύναμις κατὰ κίνησιν? Ἐνέργεια κατὰ κίνησιν—what could that be?

§ 8. *A subgroup of two metaphorical meanings:* δύναμις *with regard to the geometrical;* δυνατόν *and* ἀδύνατον *with regard to assertion*

How does this preparatory inquiry of Aristotle concerning the δύναμις κατὰ κίνησιν now look? For the Greeks, δύναμις is used in its

§ 8. Two metaphorical meanings

ordinary meaning when one speaks, for example, of δυνάμεις καὶ πεζὴ καὶ ἱππικὴ καὶ ναυτική, of military forces (either by foot, on horse, or by sea), or when one talks about the healing power of a plant, or in the expressions δύνασθαι (to have a power) and ὑπὲρ δύναμιν (beyond the power); this is the primary understanding of δύναμις. The discussion begins at Θ 1, 1046a4-11:

> ὅτι μὲν οὖν πολλαχῶς λέγεται ἡ δύναμις καὶ τὸ δύνασθαι, διώρισται ἡμῖν ἐν ἄλλοις. τούτων δ'ὅσαι μὲν ὁμωνύμως λέγονται [no comma, H.] καθάπερ ἐν γεωμετρίᾳ· [here a semicolon, H.] καὶ δυνατὰ καὶ ἀδύνατα λέγομεν τῷ εἶναί πως ἢ μὴ εἶναι. ὅσαι δὲ πρὸς τὸ αὐτὸ εἶδος, πᾶσαι ἀρχαί τινές εἰσι, καὶ πρὸς πρώτην μίαν λέγονται, ἥ ἐστιν ἀρχὴ μεταβολῆς ἐν ἄλλῳ ἢ ᾗ ἄλλο.

"That 'force' and 'to have a power' (to be capable, to be in a position to) are said (understood) in many ways, this we have demonstrated elsewhere. Among these (many ways) we shall (now) disregard those which are so designated simply according to their nominal identity. For some (meanings of δύναμις) are said in this way according to a certain identity, as in geometry; we also speak (in the sense of a certain nominal identity) of being powerful or powerless, to the extent that something is or is not in a certain manner. And yet all the meanings of 'force,' which are so understood by referring back to one and the same aspect, all have the character of something like an origin which rules over and reaches out, and are (therefore) addressed by referring back to the first way of being a force (or an origin). This first way means: being an origin of change (a ruling over and reaching out for change) in another or to the extent that it is another."

Aristotle begins with a remark on the ambiguity of δύναμις (which had already been dealt with elsewhere—Δ 12). Then the two main groups are divided: ὅσαι μέν (a6)—ὅσαι δέ (a9). The first group comprises the meanings which are so called because of a certain nominal identity. The second group pertains to the meanings of δύναμις which are connected in an appropriate and unitary way because they hold one and the same basic meaning.

The first group shall be left aside for the inquiry in Book Θ. Therefore we, too, want to pursue this only to the extent necessary to see what is thereby excluded from this inquiry into δύναμις κατὰ κίνησιν. Briefly stated, what is to be excluded are those meanings which belong to δύναμις ὁμωνύμως λεγομένη, to the meaning of force which is stated in the sense of nominal identity. Aristotle cannot mean here ὁμώνυμα ἀπὸ τύχης, that is, meanings which are designated only accidentally by the same word but which otherwise have, in terms of their content, nothing at all in common. On the contrary: ὁμοιότητί τινι λέγονται—on the basis of a certain identity, a resemblance, namely in the matter. But despite this it is once again not the relation of correspondence which essentially holds the various meanings of the second group together. Or stated more cautiously, it is not the kind of "analogy" which we have heretofore been acquainted with (πρὸς πρῶτον ἕν, cf. p. 33ff.).

Aristotle knows still another form of analogy, although he does not differentiate between the two forms with a specific designation. This was introduced later by medieval scholasticism, which calls the one form of analogy which we already know *analogia attributionis*—correspondence in the manner of an allocation to a first guiding meaning. An example of this would be "healthy." The other is the *analogia proportionalitatis*—correspondence in the manner of a likeness of proportion; for this, see *Nic. Eth.* E 6, 1131a10ff. (concerning δίκαιον, what is just). Or see *Nic. Eth.* A 4, 1096b28f.: ὡς γὰρ ἐν σώματι ὄψις, ἐν ψυχῇ νοῦς. "As vision is to the bodily eyes, so (correspondingly) is mental perception (reason) to the eyes of the soul, (ὄμμα τῆς ψυχῆς, eyes of the soul, *Nic. Eth.* Z 13, 1144a30). Accordingly, in the correspondence a *transfer* occurs from the proportionality between the eyes and vision in the physical onto the proportionality in the mental—a transfer: a μεταφορά; every "metaphor" is an "analogy" (but not in the sense of an *analogia attributionis*). Eye and eye mean here something different, but this is by no means a mere accidental and unfounded identity of the name, but rather a certain correspondence (ὁμοιότης τις) in the matter [*Sache*].

It is in this sense that δύναμις is being used here, and Aristotle refers to such a meaning in the first group (*Met.* Θ 1, 1046a6-8). It should be noted that he speaks in the plural: ὅσαι μέν, ἔνιαι γάρ—under the meanings used ὁμωνύμως, in a certain nominal identity,

§ 8. Two metaphorical meanings

there is a plurality. Plurality: this implies that δυνατόν and ἀδύνατον, which are derived from δύναμις, are also used in a multiple sense—that is, not only in geometry but rather also in a wider, more encompassing region, one which includes geometry.

Back to the text. Already by translating I have indicated how the text is to be understood. Considered extrinsically: the comma before καθάπερ must be crossed out, and a semicolon is to be placed after γεωμετρίᾳ instead; Schwegler already read this passage thus, even if he did so without giving a specific reason.[1] By doing this we achieve what is demanded by the matter itself, namely, two sentences; the first says (a7): in geometry δύναμις is used metaphorically; the second says (a8): δυνατά and ἀδύνατα are also used metaphorically. Not only does the matter which is to be treated later require that it be read this way; we clearly have in the parallel discussion at Δ 12 the same division, only in the reverse order. Δ 12, 1019b33/34 corresponds to the sentence at Θ 1, 1046a7. Here it clearly states: κατὰ μεταφορὰν δὲ ἡ ἐν τῇ γεωμετρίᾳ λέγεται δύναμις; 1046a8 corresponds to the thorough discussion in Δ 12, 1049b23-33. Both subgroups of δύναμις ὁμωνύμως λεγομένη and the accompanying δυνατόν are summed up in Δ 12, 1019b34f.: ταῦτα μὲν οὖν τὰ δυνατὰ—the δυνατόν of δύναμις qua power also belongs to this, see below—οὐ κατὰ δύναμιν, to fill in: τὴν κατὰ κίνησιν λεγομένην. What is now concretely meant by these subgroups of δύναμις οὐ κατὰ δύναμιν which differ in this way but which are all at the same time excluded from δύναμις κατὰ κίνησιν?

The group which is mentioned first in Θ 1 (1046a7) is the genuine metaphorical meaning of δύναμις, namely (in Latin) *potentia*, "power," the power of a number, for example 3 squared (3 x 3). And in fact, in Greek mathematics it is not the arithmetic proportions which are so designated (9 is the power of 3, for instance) but rather the geometrical proportions. According to tradition, this usage of δύναμις was supposedly first introduced by Hippocrates of Chios (around the middle of the fifth century; not the physician). A square constructed over a certain length and in keeping with this length is the δύναμις of this length. The

---

1. Vol. 4, p. 157. [See Editor's Epilogue for source.]

δύναμις as power is the square; thus $3^2 = 3$ squared. Accordingly, the ὅλη δύναμις of the hypotenuse in a right triangle equals the δύναμις of the other sides. What led to this meaning of δύναμις is neither clarified nor supported by evidence. We could assume that δύναμις is here named as that for which a length has the power out of itself and for itself, that is, what a length is capable of, what it yields out of itself for the construction of a geometrical figure, a spatial form; δύναμις here means what can be done with something in the broadest sense, which is not for this reason insignificant. (Plato, as well, already knew of this meaning of δύναμις in the sense of a square, as in *Rep.* 587d and *Tim.* 31, also *Theat.* 147d.)[2]

For our purposes, it is important only to see why this meaning of δύναμις is excluded from the discussion, namely, because it is *not* κατὰ κίνησιν. It is not κατὰ κίνησιν because it cannot be according to its essence. Here it is a matter of lines and spatial forms, of γραμμαί and σχήματα; these, however, according to *Phys.* B 2 (193b22ff.) are: χωριστὰ κινήσεως, ἄνευ κινήσεως—without movement, and therefore also without rest. They are completely outside movement and rest.

This applies also to the second subgroup of δύναμις ὁμωνύμως λεγομένη, and of the accompanying δυνατὸν καὶ ἀδύνατον. What is meant by this is said at *Met.* Θ 1, 1046a8: καὶ δυνατὰ καὶ ἀδύνατα λέγομεν τῷ εἶναί πως ἢ μὴ εἶναι. "We also speak this way of 'powerful' and 'powerless' to the extent that something is or is not in a certain manner"; thus with reference to certain being or non-being. To be sure, this short sentence taken in itself is not understandable. We turn once again to the parallel treatment at Δ 12 for help. Here Aristotle gives an example of what he means, and in fact one from geometry; of course, this may not be taken as though the enigmatic meaning of δυνατόν—ἀδύνατον can likewise be restricted to the geometrical and the mathematical, as is the case with the concept of "exponential power"; this is not the case, and for this reason the καθάπερ ἐν γεωμετρίᾳ may not be connected to δυνατὰ καὶ ἀδύνατα, as is so often done.

---

2. See Ross, vol. I, p. 322. [See Editor's Epilogue for source.]

§ 8. Two metaphorical meanings

At this point we know only that the δυνατόν—ἀδύνατον, and so the accompanying concept of δύναμις, is not κατὰ κίνησιν, but neither is it κατὰ τὰ μαθηματικά, τὰ ἀκίνητα καθ' αὑτά. And so the question is raised: κατὰ τί τὸ δυνατὸν λέγεται—how are we to understand the meaning of "powerful" cited above?

We find the answer through an interpretation of the lengthier presentation at Θ 12, 1019b29-30:

ἀδύνατον μὲν οὗ τὸ ἐναντίον ἐξ ἀνάγκης ἀληθές, οἷον τὸ τὴν διάμετρον σύμμετρον εἶναι ἀδύνατον, ὅτι ψεῦδος τὸ τοιοῦτον, οὗ τὸ ἐναντίον οὐ μόνον ἀληθὲς ἀλλὰ καὶ ἀνάγκη ἀσύμμετρον εἶναι· τὸ ἄρα σύμμετρον οὐ μόνον ψεῦδος ἀλλὰ καὶ ἐξ ἀνάγκης ψεῦδος. τὸ δ'ἐναντίον τούτῳ, τὸ δυνατόν, ὅταν μὴ ἀναγκαῖον ᾖ τὸ ἐναντίον ψεῦδος εἶναι, οἷον τὸ καθῆσθαι ἄνθρωτον δυνατόν· οὐ γὰρ ἐξ ἀνάγκης τὸ μὴ καθῆσθαι ψεῦδος.

A translation which is at the same time an explanation: "Powerless means here that whose opposite necessarily is what it is as it is manifest; for example, the diagonal of a square is powerless to have the same measure as the side of the square; we speak of a being powerless because such a thing—having the same measure as the side of the square—conceals, that is, it conceals the diagonal in its own commensurability; for it is not only directly manifest that it is, on the contrary, incommensurable with the side of the square, but rather it is manifest that it is necessarily incommensurable in this way. The determination of the commensurability by the side of the square is not only misleading, not only a concealing of the matter, but it conceals out of necessity. But then the opposite to this, to being powerless in this sense, namely being powerful for . . . , emerges when the opposing determination does not of necessity conceal; so, for example, a human who is now in fact standing has the power to sit; for the determination 'not sitting' does not necessarily conceal the 'what' that the human is."

What is being said here? Two examples; one for the ἀδύνατον, one for the δυνατόν. The first example comes from geometry (the diagonal); the second example is drawn from the field of beings which are in the widest sense experientially accessible and present (an encountering human). More exactly, however, two sentences, two *assertions*, are drawn upon as examples. The first assertion states that the diagonal has the same measure as the side of the square (can be measured by this); the other assertion states that this human is sitting here. We can infer from this—negatively: that the enigmatic meaning of δυνατόν and ἀδύνατον is not restricted to geometrical-mathematical relations; positively: this meaning is somehow related to the character of an assertion of something about something, the ἀπόφανσις. This already indicates that a thoroughgoing thematization of the meaning of δυνατόν and ἀδύνατον κατὰ τὴν ἀπόφανσιν would demand an extensive discussion of ἀπόφανσις and of λόγος in general—concerning this, cf. especially *De interp*. 12 and 13—but this exceeds the limits of the inquiry into δύναμις κατὰ κίνησιν. Aristotle wants to say nothing else with the exclusion of δύναμις ὁμωνύμως λεγομένη.

And yet we have to arrive at an understanding of one thing: in what sense, as well as for what reasons, we can speak of a δύναμις, and thereby also of δυνατόν and ἀδύνατον, precisely in the realm of ἀπόφανσις. Only with respect to an explanation of this context do I offer a brief interpretation of the passage of Δ 12. This is not the occasion to deal fully with this text in all its essential respects.

Aristotle states: The diagonal is powerless to have the same measure as the side of a square—σύμμετρον εἶναι. This is ψεῦδος—it distorts and conceals what the diagonal manifestly is. If we state the commensurability of the diagonal in terms of the side of the square, then we do not allow the diagonal to be spoken of with regard to what the diagonal itself tells us. And what does it say? This means: What is it itself? The diagonal denies the saying of its commensurability to the assertion concerning it. It denies and forbids this, because in this regard the diagonal itself denies the attempted measurement by the side of the square; it is inherently without the power for such measurement. It would not be compatible with it. It is powerless, not allowing of such a thing; that is, with respect to being measured by the same standard, the diagonal is incompatible with the sides. Ἀδύνατον, being without the power for something, now means: failing in something, not being compatible with this something, with something, namely, which might be attributed in an assertion. That which in its "what," according to its inherent content, fails in something in this way and cannot bear it, must deny (forbid) the assertion

§ 8. Two metaphorical meanings

as something able to be asserted. The diagonal, in accordance with what it is itself, denies, in that it would not bear the measurement by the side. Hence it denies the attribution of commensurability to the assertion about it. —But now if the opposite of that in which it fails and which it denies (forbids) is attributed to it—namely, the opposite of commensurability, the incommensurability with the side of the square—then this opposing attribution is that with which the diagonal would be compatible, is even that upon which it insists. What is attributed in this way, therefore, says something which the diagonal makes manifest in being what it is: ἀδύνατον μὲν οὐ τὸ ἐναντίον ἐξ ἀνάγκης ἀληθές.

In what follows, we need to pay attention to the perspective issued in by this meaning of δύναμις and why this meaning is excluded at least for now from the discussion in Book Θ, even though, as we shall see, it is one meaning of being.

*

Aristotle begins his inquiry into δύναμις and ἐνέργεια with the discussion of δύναμις κατὰ κίνησιν. This discussion itself was introduced through the distinction between two main groups of meanings of the word δύναμις; we have two accounts of this, θ 1 and Δ 12. The second group comprises nothing less than δύναμις κατὰ κίνησιν, which is to be our theme. The first main group is introduced only in order to be excluded. It is excluded because here δύναμις does not function κατὰ κίνησιν. The treatment of this excluded group states negatively what δύναμις κατὰ κίνησιν is, and so indeed achieves something for the clarification of our theme. This first group itself is constituted of two meanings of δύναμις, both of which—considered in terms of the second group—are used metaphorically: (1) δύναμις κατὰ τὰ μαθηματικά—"exponential power"; (2) δύναμις qua δυνατόν (τῷ εἶναί πως). Regarding (2): The account proceeds from ἀδύνατον, with the example of the incommensurability of the diagonal with the side of the square. Ἀδύνατον—"powerless" means: not to bear, to deny and so to forbid. The denial is the forbidding of an assertion—or else the demand for its opposite.

Δυνατόν is now to be understood accordingly. What has the power for something, does not deny and is able to bear it. The human who is standing—on the basis of what this human is as such—by no means fails to exist also as one who also sits. The human being does not deny, does not forbid that being seated be attributed to it. This determination is compatible with it.

Δυνατόν and ἀδύνατον mean then a non-denial and a denial, a being incompatible and a being compatible, which means, however, a non-concomitance and a concomitance with . . . , a non-togetherness and a togetherness (σύν-θεσις) as presence with another or non-presence—which in the Greek sense means—a certain being or non-being (of something in unity with something else). Καὶ δυνατὰ καὶ ἀδύνατα λέγομεν τῷ εἶναί πως ἢ μὴ εἶναι (Θ 1, 1046a8). "We also speak of being powerful and being powerless, to the extent that something is or is not in a certain manner." We encounter this being and non-being as compatibility (δυνατόν) and incompatibility (ἀδύνατον) most immediately and almost tangibly in the assertion that something is such and such or is not such and such. With regard to the assertion the δυνατόν is ἀπόφανσις. Here we have the meaning of δύναμις—we may say—κατὰ τὴν ἀπόφανσις. It belongs to the essence of this, however, to be able either to uncover or to conceal: ἀληθές or ψεῦδος. (We are already acquainted with this as a basic way of being.)

Here, then, is the δύναμις which pertains to the φάσις, to the saying, the dictum. From this we surmise that here, with the clarification of this meaning δυνατόν and ἀδύνατον, we are dealing with an ἐναντίον—which, not accidentally, is found directly in the definition of ἀδύνατον—as well as with the ἀντί—that which lies over and against—and with the φάσις, as ἀντίφασις: the saying, the dictum as counterdiction and contradiction. It is for this reason that we find ἀδύνατον in the so-called principle of contradiction. Γ 3, 1005b29f.: ἀδύνατον ἅμα ὑπολαμβάνειν τὸν αὐτὸν εἶναι καὶ μὴ εἶναι τὸ αὐτό—the same speaking and understanding human, as itself, stands powerless, cannot tolerate or permit, with reference to one and the same being, that this being simultaneously be taken in advance as being and not being. Whoever understands this ἀδύνατον from out of its ground, and does not just simply continue to prattle on about

§ 8. Two metaphorical meanings 55

it, as is so often the case in logic and dialectics when it comes to the so-called principle of contradiction, this one has grasped the basic question of philosophy. But that way is a longer and more arduous path. This path is precluded from the very beginning if it is maintained that Aristotle's principle of contradiction is not only logical but ontological as well. Aristotle knew neither the one nor the other. His posing of the question lies before this ossification into scholastic distinctions. It is no less erroneous to speak of logical possibilities, if this is supposed in any way to mean a formal freedom from contradiction—what is meant here by contradiction is much more contained in the matter itself. The way toward an understanding of the so-called principle of contradiction must first traverse an overall understanding of δύναμις in all its dimensions.

Now if Aristotle excludes from his inquiry the discussion of δύναμις and δυνατόν in the sense we have thematized here, that does not at all mean that the excluded meaning is fundamentally devoid of relation to the question of δύναμις and ἐνέργεια; just the opposite holds.

We who are of the modern age are not yet at all prepared for an effective interpretation of this passage at Δ 12. There is only one additional sentence here to be adduced as external evidence, 1019b30ff.: τὸ μὲν οὖν δυνατὸν ἕνα μὲν τρόπον, ὥσπερ εἴρηται, τὸ μὴ ἐξ ἀνάγκης ψεῦδος [εἶναι, H.] σημαίνει, ἕνα δὲ τὸ ἀληθὲς εἶναι, ἕνα δὲ τὸ ἐνδεχόμενον ἀληθὲς εἶναι. "Being powerful, having the capacity for something, means on the one hand being-not-necessarily-concealing, on the other hand being-revealing, and then again it means being in a position to have the capacity in the sense of being-revealing." We see only very roughly that δύναμις is here καθ'ἀλήθειαν, καθ' ἀληθὲς ἢ ψεῦδος. I happened to recall quite incidentally that the ὄν is stated, in another sense, κατὰ δύναμιν and then also ὡς ἀληθές; thus ἀλήθεια, εἶναι, and δύναμις move closer together.

It deserves to be asked why precisely a metaphorical meaning of δύναμις and δυνατόν, in the sense of compatibility and incompatibility, arose in reference to mathematical objects, and why at all in reference to being true and being not true—an event of the greatest consequence for the basic questions of philosophy, above all in modernity (Leibniz and Kant—possibility as the lack of contradiction and compatibility). The question concerning the intrinsically determinative ground of the concept of δύναμις in its relation to truth is all the more pressing in that this metaphorical meaning of δύναμις remains manifestly connected—although in an admittedly obscure way—with the proper meaning. Or does the δύναμις ἐπὶ πλέον λεγομένη (cf. above, p. 41) first clear a path for comprehending the connections which have just been put into question?

Aristotle ends by stating 1019b34f.: ταῦτα, that is, that which has been addressed as δυνατά in the way mentioned—ταῦτα μὲν οὖν τὰ δυνατά οὐ κατὰ δύναμιν—is not "with a view to δύναμις," namely in the usual and proper sense as δύναμις κατὰ κίνησιν, so that to complete it: οὐ κατὰ δύναμιν τὴν κατὰ κίνησιν. Δύναμις, force and having force for . . . , is instead carried over from κίνησις, as the genuine dwelling place of its meaning, to ἀλήθεια—as was demonstrated quite unambiguously with the example of the diagonal: κατὰ μεταφορὰν δὲ ἡ ἐν τῇ γεωμετρίᾳ λέγεται δύναμις.

§ 9. *The guiding meaning of* δύναμις κατὰ κίνησιν

Our treatise, Book Θ, excludes the metaphorical meaning of δύναμις. And what happens with the usual, original, and proper meaning of δύναμις, force? It too exhibits multiple meanings. Yet these various meanings are no random collection but are all understood πρὸς τὸ αὐτὸ εἶδος (Θ 1, 1046a9)—with reference to the same outward appearance. Here again we meet the πρός (cf. above, p. 33)—in distinction from κατά, which, for the most part, means inclusion under a genus, γένος, or species, εἶδος. The meaning of our passage is completely missed if we take εἶδος for "species." That would imply that the ways of δύναμις κατὰ κίνησιν to be discussed in what follows are subspecies of a higher type. This, however is not the case, but we have instead once again a relation of analogy.

The corresponding ways of δύναμις all coalesce in this: they are all ἀρχαί τινες (a9). They are all like that from which something proceeds. These multiple meanings of δύναμις correspond to each other

§ 9. *Meaning of* δύναμις κατὰ κίνησιν 57

in that they all in their meaning as ἀρχαί go back to a first ἀρχή, back to a meaning of δύναμις which comes into play before all the others. This πρῶτον ἕν πρὸς ὅ (cf. a10), this first one back upon which all the corresponding meanings are understood, we shall call the meaning which guides all the correspondence, or the *guiding meaning;* cf. Δ 12, 1020a4: ὁ κύριος ὅρος—the dominant meaning.

This guiding meaning says (Θ 1, 1046a10f.): being the origin of change, an origin which as such is in a being other than the one which is itself changing, or, if the originary being and the changing are the same, then they are so each in a different respect. Δ 12 has the formulation: ἀρχὴ κινήσεως ἢ μεταβολῆς ἡ ἐν ἑτέρῳ ἢ ᾗ ἕτερον. From these formulations of the guiding meaning of δύναμις, force, we now surmise the following:

(1) Force is ἀρχή—origin of . . . (2) Of what? Of a change, a movement: κίνησις. (3) The origin of the change is in something other than the change, which means in a being that is not the same as the one that changes. (4) We find the added phrase ἢ ᾗ ἄλλο: or (however) inasmuch as that within which the change is brought about is the same being as that which brings about this change, (then) this happens only in such a way that the ταυτόν here is that which in one respect is changing, and in another respect that which brings about the change.

The beginning of Δ 12 offers the example: οἷον ἡ οἰκοδομικὴ δύναμις ἐστιν ἣ οὐχ ὑπάρχει ἐν τῷ οἰκοδομουμένῳ—so is, for example, the οἰκοδομικὴ τέχνη, the art of building such a δύναμις, that is, something from out of which . . . , something by virtue of which a change in the stones, bricks, and wood succeeds in becoming a house. The art of building as ἀρχή is itself not present in the built house. And this is always the case when δύναμις is used in the way indicated. As source, as ἀρχή, it is ἐν ἄλλῳ—in another. This expression, ἐν ἄλλῳ, is not originally to be related to μεταβολή (cf. below, p. 72f.). Ἀλλ' ἡ ἰατρικὴ δύναμις οὖσα ὑπάρχοι ἂν ἐν τῷ ἰατρευομένῳ. "But the art of doctoring, although it is a δύναμις, may nonetheless be present occasionally in that which is itself being doctored," namely when the physician treats himself as the one who is sick. Here the ἀρχή is that from out of which the change from sickness to health originates—οὐκ ἐν ἄλλῳ—not in another but instead ἐν ταὐτῷ, in one and the same being. This is so, however, to the extent that the doctor, being a doctor, is something other—ἢ ἄλλο—than being sick. This being sick does not occur in the doctor insofar as he is a doctor but insofar as he is a human, a living being having a body. As a doctor the human cannot lack anything in the sense of being sick; but as a doctor the human may very well lack something, if the doctor is a quack.

Already the point of departure of the Aristotelian inquiry into δύναμις κατὰ κίνησιν shows that he is not after a mere collecting of word meanings in order to count them up one after the other; his business is not "lexicography," but from the very start he is aiming for an understanding of the matter in view. And this determination of the "essence" of the matter is again done not for the purpose of establishing a usable "terminology" and an academic parlance but rather to make visible at once the manifoldness of the essence and its possible modifications. Through this delimitation of the guiding meaning, we are placed from the very outset into the realm of a questioning about this essence. Expressions such as ἀρχή, κίνησις, ἄλλο, ἕτερον, ᾗ ἄλλο, ᾗ ἕτερον point to essential moments. Admittedly—the primary, guiding meaning does not at all permit that it now be dealt with on its own in a detached manner, but rather requires entering into the whole accompanying and corresponding nexus of the matter which is guided by and subordinated to it.

Aristotle himself refers to another treatment of the same question concerning the essence of δύναμις κατὰ κίνησιν, in Δ 12. At first one finds no difference between the two accounts. Θ 1 is more concise, Δ 12 broader and aided by clarifying examples. Nonetheless a very definite intent is emphasized several times in Θ 1: namely, to show how the delimitation of δύναμις which guides all the correspondences is somehow already co-present in all these varying meanings—ἔν-εστι, ἔν-υπάρχουσί πως (1046a15 and 18). Thus we shall take both accounts together in such a way that we shall rely thematically on Θ 1 but take from Δ 12 above all the elucidating examples and special features.

§ 9. *Meaning of* δύναμις κατὰ κίνησιν

a) Approaches to the phenomenon of force and a rejection of the so-called transference

Yet before we proceed, we want to make our understanding of the matter being treated here under the title δύναμις κατὰ κίνησιν a bit more lively and concrete so that we might gain a sharper view of the uniqueness of the Aristotelian approach and method. But not only this. It is much more important that we first of all prepare ourselves in order to be able to experience that it is not simply a game of thoughts and concepts which are playing themselves out in the text—without resistance, without home and need—but that here, as in every actual philosophy, the power of a Dasein is pressing forward toward the freedom of the world, and that this philosophizing is still *here*, not here in the impoverished presence of a supposed Aristotelianism but here as an indissolvable bond and an unending obligation.

Nonetheless, with such an attempt we run directly into a totally untraversable area fraught with entangling connections that have long since been expressed in language but totally deprived of conceptual thoroughness. The usage of language is accordingly now a matter of changing feelings and tastes. Viewed in this manner, it appears overbearing to lay claim to definite expressions for definite meanings.

We need first of all only to remember what we ourselves have already undertaken in this regard. With the translation of the passage relating to the incommensurability of the diagonal with the side, I expressed ἀδύνατον using the German *unkräftig* [powerless], a word which is surely odd in this context. This was done intentionally in order to retain the correspondence to the word δύναμις—*Kraft* [force or power]. The diagonal is powerless to do something; we would like to improve upon this by choosing "lacking the capacity," and so replace force with capacity [*Fähigkeit*]. Yet neither does this exactly hit the mark: the diagonal lacks the capacity to be measured. Capacity—we think first of all of a capacity to accomplish, of a making do, even if only by bearing something or putting up with something; so, for example, we talk about the load capacity of a bridge. In no way does the diagonal have a capacity for something in this sense. And then again, in the discussion of μεταβολὴ ἐν ἄλλο ἢ ᾗ ἄλλο (of the change in another or to the extent that it is another), namely with the distinction between the production of a house through building and the production of one's own health through the activity of a doctor, we spoke of the art of building [*Bau-kunst*], the art of healing; in this last case the δύναμις is not a healing power [*Heil-kraft*]. A doctor does not have any healing power, as does a plant or a medicinal herb, but instead the doctor possesses the art of healing, and the builder the art of building; here we are taking δύναμις as art or as ability.

Someone can play the violin; by this we do not mean only that he has attained this ability but that he has cultivated a capacity—was able to cultivate this capacity because he already had it. This having of a capacity we understand as being talented: δύναμις as talent. We do not say, of course, that the railroad bridge is talented in bearing the heaviest trains. In contrast, we do speak of a talented person, and so of a person of capacity. One who has capacity is enabled [*befähigt*]. Although one who has the capacity to be a good teacher is not thereby competent [*Befähigung*] in the sense of being qualified. In contrast, we call the power and the capacity of sight in the eye the faculty of vision [*Sehvermögen*]. We say powers of the soul, faculties of the soul, but never capacities of the soul; at the most we say psychic capacities [*seelische Fähigkeiten*]. Again, we do not speak of the sovereign force [*Herrscherkraft*] of a king but of his sovereign might [*Herrschergewalt*]. And the violent force which a brutal person might exercise we distinguish from the power [*Macht*] of an idea; yet on the other hand we call brutes despots [*Machthaber*]—despots because they do not have power and cannot use it. Instead they abuse it in the extreme because they possess only the means to employ violence.

We could continue in this manner. Only to exhibit the multiple usage of the word? Only to demonstrate that we use different words at the same time for the same thing? No, just the opposite, in order to see precisely that our use of words such as force [*Kraft*], capacity [*Fähigkeit*], art [*Kunst*], talent [*Begabung*], capability [*Vermögen*], competence [*Befähigung*], aptitude [*Eignung*], skill [*Geschicklichkeit*], violent force [*Gewalt*], and power [*Macht*] is not completely arbitrary;

§ 9. Meaning of δύναμις κατὰ κίνησιν

and even when we substitute one for the other (for example: the power of sight, ability to see, faculty of vision), we still hear differences in this. We understand at once certain differences in each respective case. And yet how helpless we are if asked directly: What do you understand by power? What is called force? What does aptitude mean? Is it even possible in this way to define these expressions simply off the cuff, without further ado? Can the matter at stake in these words be grasped at all in the same way as the knife on the table or the book on the bench seat? And if not, what then is the point of searching? To find the realm within which what has been so designated can be determined!

But to where does all of this lead? Let us leave such indeterminate, undecided, fleeting, and polymorphic things to language! What would be the point of a sum of fixed definitions with words grafted onto them and thereby made unequivocal? That of course would be the decline and death of language. And yet what is at issue here is not language as such, nor is it words and their meanings. We want instead to discern slowly that in the string of words tallied off something is meant, something which in a certain manner is the same, even while being different. All this multiplicity—is it something arbitrary and trivial, or does a basic occurrence of every being and of each way of being here present itself to us? Force—the forces of material nature; what would nature be without forces? Capacity—the capacities of a living being; capability—this and that capability of the human; art—the art of Michelangelo, of van Gogh, what would we understand of both if we did not understand art? Violent force—the violent force of Napoleon; power—the power of the divine, the power of faith.

One might be tempted to say that running through all of these is ability. Thus what is at issue here are specific kinds of ability, and ability is the general concept under which these other types fall. And what is this ability? This is something utmost and does not permit of being defined further. With this, philosophy is finished. It remains only to be said that according to the modern position of science, it is of course pure mythology to speak of the forces of nature or of the capacities possessed by the bee, or the faculties of the human soul. These are "naive hypostatizations" whose origins have long since been discovered: namely, the human transfers onto the things outside inner experience, where something like accomplishing and the ability to accomplish is encountered. Subjective experiences in the internal soul are projected and transferred outward to the objects. It is said that Aristotle's conceptual pair, δύναμις—ἐνέργεια, has to be explained from this perspective as well. It is a conceptual schema which owes its origin to a naive world view and hence is applied by Aristotle everywhere uncritically. One speaks of the conceptual pair δύναμις—ἐνέργεια as a "universal means" with which Aristotle sought to resolve all questions.

It will become apparent how all this stands up. In what follows we will make clear to ourselves in eight points that this kind of explanation of the origin of the concept of force in subjective experience not only is untenable but even tends to push the questioning in a direction which ultimately shirks from the actual problem. On this account it is no accident that today, despite the long tradition of this conceptual pair, we do not have even the slightest serious effort in philosophy to press in on the phenomena which lie behind this title δύναμις.

*

The inquiry in Θ 1 begins with the discussion of δύναμις in its usual meaning; the higher philosophical meaning is set aside, or more exactly, it is not even yet known but must first be exposed. Δύναμις in its usual meaning is δύναμις κατὰ κίνησιν. To a first but excluded subgroup belong those variations which are used in a transferred or metaphorical sense of compatibility or incompatibility. These have (a) a mathematical sense (power) and (b) a logical sense ("logical" because related to λόγος, assertion). Moreover, we can infer that the meaning in the higher, philosophical sense does not concur, say, with its "logical" sense and so must mean still something else. It remains open how the two are connected. For now the topic is δύναμις κατὰ κίνησιν in its usual and proper meaning. This reveals again a plurality of meanings, but πρὸς τὸ αὐτὸ εἶδος. With this guiding meaning, which is one and the same, we insisted upon four points: (1) ἀρχή; (2) μεταβολή; (3) ἐν ἄλλῳ; (4) ᾗ ᾗ ἄλλο. —Independently of the

§ 9. Meaning of δύναμις κατὰ κίνησιν

interpretation and in order to push forward toward an understanding of the matter itself, we called to mind phenomena which are to be found under the title δύναμις and which we designated as power, capacity, competence, proficiency, aptitude, talent, skill, being accomplished, capability, power, violent force. This is no groundless rummaging around in word meanings. In these expressions there lies a certain ordered relation to certain realms of being, and we see that they provide the basic determinations for these realms; without them we would be utterly unable to comprehend such realms. Formally and abstractly, one could gather all these phenomena together as "ability." This comes to us from an "experience" of ourselves: we experience in ourselves ability or inability. And so, the origin of the concept of power lies here, in a subjective experience. From there it is transferred onto things, "sympathetically"—and if viewed in a strict scientific manner, without warrant. How do things stand with this current explanation of the origin of the concept and essence of force?

We are asking: Is anything actually explained by referring the positing of forces in things and objects themselves back to a transferral of subjective experiences into the objects? Or is this popular explanation a sham; namely, is it something which for its part is in need of explanation in all respects and, when explained, untenable? The said explanation is indeed a sham. We shall try now, with attention to what comes later, simply to become familiar with this by adducing a few guiding thoughts. From this it shall become apparent how the said explanation fails to recognize its own presuppositions.

1. The stated explanation presupposes as self-evident that what transpires in the inwardness of subjectivity is more easily and more surely comprehended than what we encounter externally as object.

2. It is assumed that the subject, the proper I, is that very thing which is first of all experienced and which thereby presents itself at any time as the nearest. From this is derived what undergoes the transferal onto the objects.

3. The said explanation neglects to demonstrate why such a transferal from subjective determinations onto the objects is carried out at all.

4. In particular it fails to ask whether the objects themselves do not, after all, demand such a transfer of subjective experiences onto them.

5. If there exists such a demand, and if it is not pure arbitrariness that we, for example, name one landscape cheerful and another melancholy, then it must be asked how the objects themselves are given prior to the metaphorical, transferred comprehension and the sympathy of such a mood. What is their character as objects such that they demand such a transfer?

6. It is not taken into consideration that, if the objects themselves in accordance with their intrinsic content and their way of being require such a transfer in order to be addressed, for example, as forces and powers, then indeed a transfer is not needed in the first place; for in this case we would already find in them what we would attribute to them.

7. Recklessly explaining certain objective thing-contents—for example, real forces and efficacious or effective connections and capacities—as subjective transferals results in even those forces, capacities, and capabilities peculiar to subjects as such being misconstrued in their own proper essence.

8. Because of this and on the basis of all the said shortcomings, the way to a decisive question remains closed off, and this question runs thus: In the end, is what we are here calling force, capacity, etc., something which in its essence is neither subjective nor objective? But if neither the one nor the other, where then do these phenomena belong? Do they at all allow themselves to be determined from out of an origin? But then what kind of explanation is such a determination of origin?

The difficulty of comprehending the essence of the phenomena designated under the heading "force" does not simply lie in the peculiar content of these phenomena but rather in the indefiniteness and ungraspability of the dimension in which they are properly rooted. To be sure, such a comprehension cannot be achieved through wild speculation and dialectic.

b) The apparent self-evidence of causality and the Aristotelian essential delimitation of force

We will never come any closer to these phenomena as long as we do not first attempt an initial interpretation of the phenomena now under discussion, free from crude prejudices such as those just mentioned,

§ 9. *Meaning of* δύναμις κατὰ κίνησιν

even if such an attempt runs the risk of not taking us far enough. Because there is not an active understanding of these questions, there is a total lack of appreciation for what Aristotle was the first to achieve in this regard. We who have long since become too clever and all too knowing have lost the simple ability to detect the greatness and the accomplishment of an actual engagement and undertaking. All too well versed in the commonness of what is multiple and entangled, we are no longer capable of experiencing the strangeness that carries with it all that is simple. How are we supposed to receive and even appreciate what Aristotle has to offer us?

We will succeed in this most readily if we give ourselves over for a time to what we believe primarily to be the way in which we come across phenomena such as force and the like. If we want to give something such as force its due, and to make sense of it, then this thing, so it appears, must be secured in advance. How then do we discern a force? Do we discover forces as simply as we discover trees, houses, mountains, and water or the table and chair? For example, we speak of the hydraulic power of an area, of a mountain, of the illuminating power of a color, of the load capacity of a bridge, of the gravitational pull of the earth, of the reproductive capacity of an organism, and so on. Do we ever directly discern here a force on its own, that is, do we perceive it in advance? Of course not, we shall say. For we experience the load capacity of a bridge only through what it accomplishes, for example. Likewise, we comprehend the illuminating power of a color only in the effect of its lighting. And we comprehend the capability to act only by its success or failure. Forces do not allow themselves to be directly discerned. We always find only accomplishments, successes, effects. These are indeed what is tangibly actual. We come upon forces only retrospectively, and for this reason, to be sure, the positing of forces is in a special way continually subject to suspicion.

But do we find "effects" directly? The lighting of the color, is that simply an effect? Is the falling of the stone, a being drawn toward, simply an effect? By no means do we experience something immediately as an effect in distinction from a merely mediated inferring of forces. We experience something as an effect only if we take it as effected, effected by the force of something else, thus having force as its cause. If this is the case, then we must say that the experiences of forces and effects are equally mediated or equally immediate—assuming that it is clear what we mean by immediate. Effects are discovered on their own just as little as are forces. Forces are no less understandable than effects, and the latter are just as enigmatic as the former.

On the other hand—we experience both of these directly in everyday experience. No inference from the effect to the force is needed, since to experience effects already means to encounter forces. The need for retrogressive conclusions, or better, for the considerations and questions which lead back from the effect in particular to the cause, first arises only in order to achieve a more proximal determination of a cause already posited as present. But then what is encountered already stands in a relationship of cause and effect. Only in light of this relationship of causality—and this means the being-a-cause of beings and the being-caused of beings—do we find forces in being, and only in this light are we capable of measuring forces. Force is accordingly a concept which follows from causality (Kant, *Critique of Pure Reason*, A 204, B 249).

But is something now clarified in saying that force is a concept which follows from causality? In any case, with this a task is posed and there is a gesture in the direction from which we can expect an opening concerning the essence of force. The more original question is then: What does it mean *to be a cause?* Thus the question is becoming broader and more general, yet not in the least easier or more transparent. But let us for once follow up on what we are asking about.

We began with the question which presented itself: How do we discern a force at all? This yielded: We encounter forces as causes only in light of the causal relationship. Through this digression into the question of how a force is discerned, we have learned something about the essence of force after all, even without having asked explicitly and simply what force in general is. The discernibility of a force, the access to it, must evidently be co-determined by what force in itself is; correspondingly, force is co-decisive concerning its own comprehensibility. And yet have we actually gained any knowledge about the essence of force (δύναμις)? Strictly speaking, only that it is discernible in light

§ 9. Meaning of δύναμις κατὰ κίνησιν

of the causal relationship. And yet that still does not say anything about force itself. Or does it?

Causality—as being a cause of and for something—is a determination of the being of beings inasmuch as beings are in movement or else inasmuch as they can be in movement. Correspondingly, something shows itself to us as an effect at all only if it has already somehow become questionable for us from the very beginning with regard to its becoming; thus only when it has become an object for us in its becoming and being moved. Becoming questionable means here: asking about . . . in the sense of why? from where? For example, we could take the falling of the leaves in autumn as only a mere gliding downward, as just this falling and nothing more. And we could regard the rising of smoke over a farmhouse in the same way. And if this should need to be explained, then it is not all necessary to resort to cause and effect as explained conventionally. We, too, could interpret both cases, as Aristotle observed, as things going toward their place. (A possible explanation of nature which until today is not in the least refuted, in fact not even grasped.) But if we experience force as being the cause of something, then force is in itself related to being-moved and to movement, and indeed precisely as that which thereby is distinguished from its being-moved and movement, as what is not the same as these.

Now if we do not allow all that has been said to slip immediately away again or, and this is the same thing, all too readily take it as self-evident and therefore pointless, then it turns out that we have achieved something. Force has the character of being a cause; cause [*Ur-sache*]: an originary thing [*Sache*] which allows a springing forth, that from out of which something is, namely as a particular being-moved, and this again in the form that this being-moved is in its movement a different thing from the cause. The insights we gained here into the essence of force via the circuitous path of the inner unfolding of causality, Aristotle saw in a decisive and essential moment and brought univocally to word and concept: δύναμις is ἀρχὴ μεταβολῆς ἐν ἄλλῳ ἢ ᾗ ἄλλο—the origin of change, which origin is in a being other than the changing being itself, or, in the case where the originating being and the changing being are the same, then each is what it respectively is as a different being.

This is a meager conceptual determination in light of which we face the choice either to leave it to itself as a self-evident platitude or to take hold of it as a decisive step toward a determination of the essence of force. The first path closes itself off. We need only to take note of the concepts which come to the surface in this essential delimitation of δύναμις: ἀρχή, μεταβολή, ἄλλο, ἕτερον—the whence, alteration, otherness or difference, relation. These are pronounced basic concepts of Aristotelian philosophy and of ancient philosophy in general. How so? They point to the ultimate horizons from out of which antiquity understood and attempted to grasp the being of beings. If we now want to understand the Aristotelian determination of the essence of δύναμις, we will succeed only if we understand it still more originally. This demands tracing and securing the horizons that are designated through these basic concepts. That sounds like a self-evident demand, and yet since Aristotle it has not once been pursued. Much more, this essential delimitation of Aristotle was taken as a fixed definition, which was never actually thoroughly interrogated, nor was what lay behind it questioned. Hence it remained mute and became trivial.

Mistaking the content and the guiding role of this delimitation led to demands being placed upon it at the beginning of modern science which it neither could nor would satisfy. One declared it "scientifically" useless, whereby one understood under science: mathematico-physical research of nature. Whereupon, of course, absolutely nothing was decided philosophically. The most secure and comfortable path has always been to make something harmless and insignificant by admitting and acknowledging it once and for all to be self-evident. Thereby the established view has agreement from all sides. Such is the case with δύναμις and ἐνέργεια in the judgment passed by the history of philosophy. And so we later ones and latecomers are in a peculiar situation. We must first of all recapture for ourselves the self-evident as something worthy of question.

And so it is necessary to remain with the Aristotelian definition that has been presumed a triviality, and to set it free in its essential content. If we have even for one moment actually made this demand clear to ourselves, then we can see that the philosophy of Aristotle, and thus every philosophy, remains closed off to us as long as we do not go beyond it in the direction of its proper origins and questions. If that

§ 9. Meaning of δύναμις κατὰ κίνησιν

should occur, then what is presumptuous in our task will make itself evident of its own accord. To be sure. But then to philosophize is always nothing other than the greatest presumption ventured by human Dasein given over to itself.

Thus it is necessary to surpass Aristotle—not in a forward direction in the sense of a progression, but rather backwards in the direction of a more original unveiling of what is comprehended by him. With this we are saying further that what is at issue here is not an improvement of the definition, not a free-floating brooding over individual lifeless concepts. Rather, this going beyond which leads backwards is at once the implicit struggle by which we bring ourselves again before the actuality that prevails tacitly in the concepts that have lost life for the tradition. Whether this monstrous task succeeds or fails, that is a later concern. It is enough if we experience in this struggle only that we are too weak and too unprepared to master what has been given to us as our task. This may then at the very least awaken in us the one thing which belongs in no small way to the presumption of philosophizing and about which there is nothing more to say: the awe before the actual works of spirit.

(We are today as far removed from all of this as possible. Today we talk about the academic proletariat. One understands by this the mass of intellectual laborers who cannot be professionally accommodated. In this lies the opinion that the proletariat would be eliminated once employment opportunities were procured for these masses. The academic proletariat prevails, however, in a completely different way. One must say without exaggeration: A scientific "peak performance" today—to use that dreadful expression—has long had no need for the aristocracy of the spirit. Those who have long since been provided for are also precisely those who have long since been proletarians because they feel complacent in their impotence toward aristocracy. They have neither the scent for the height of spirit—which is struggle—nor the inner power to bring it to mastery.)

Only if we become truly humble is the scent awakened for what is great, and only if this occurs do we become capable of wonder. Wonder is, however, the overcoming of the self-evident.

We are now to occupy ourselves with the self-evident in the Aristotelian definition of δύναμις: ἀρχὴ μεταβολῆς ἐν ἄλλῳ ἢ ᾗ ἄλλο. We are to take this up and to let it hinder us from handling it comfortably and expeditiously with an air of superiority, to let it teach us that such expediency constantly threatens to ensnare us.

We find just such a lure of the self-evident here. In our preliminary discussion of the concept of force, we observed that it is connected in a certain way with causality. According to Kant, "force" is a concept which follows from "cause." Now we know that Aristotle often also uses αἰτία for ἀρχή, which we translate as "cause." We gain from this the undeniable fact: δύναμις is comprehended by Aristotle as a kind of "cause." This suggestive consideration would not be the lure which we claim it to be if what has just been said was not overwhelmingly "correct." And yet with this correct view in which Kant and Aristotle concur, where force is a kind of cause, we have already allowed ourselves to be pulled away from what Aristotle said. For what matters most in the Aristotelian delimitation of the essence of δύναμις is to see how preliminary and careful and thus how completely open it is. What is at issue here is not at all a cause-and-effect relationship, where we immediately think of the transfer of force, the effect of distance, and so on, and then puzzle over the secret relationship between cause and effect. Much more, it is maintained clearly and simply: force is an origin, the from-out-of-which for a change, and this in such a way that the origin is different from that which changes. We must bring ourselves face to face with the references mentioned here as simply as possible and without premature profundity so that through this we might come to learn how in fact what we call a force is contained therein.

*

Our purpose here is to come closer to the task of an essential determination of δύναμις in its peculiarity and difficulty. To this end we followed the popular procedure of explaining the origin of the concept of force, and so misconstrued the breadth and depth of its essence; we do not need to repeat the eight points. Until very recently it was still considered a distinctive achievement to explain scientifically

§ 9. Meaning of δύναμις κατὰ κίνησιν

the very appearance of living nature and living beings while excluding all so-called capacities and forces. Indeed, today it is still fundamentally the hidden claim of the science of biology, to make do—mechanistically—without such a presumption of forces and capacities. And this is in part as it should be, if only because the elaboration of the essence of life has still not been taken far enough if questions which point elsewhere are to be brought to fruition. Vitalism is emerging as the countermovement to mechanism; of course, this in no way guarantees the right understanding of what is basically at issue here. But this is not the place to enter into this. (Such biological viewpoints are particularly suited to giving character to a world view.) —We were left with the unavoidable result: We must for once actually attempt to determine the essence of force, and so for once to follow a certain natural path. First it is necessary to discern what something such as force is. How do we discern a force? It appears that force does not let itself be discerned immediately, only "effects" let themselves be discovered. Nietzsche, for example, also based his will to power on this thesis and so based it upon an errant foundation. Viewed more closely, effects are directly encountered just as little as are forces—or just as much. We have cause and effect simultaneously—the cause-effect relationship and, in its light, "force." Force is accordingly a derivative concept. The question of its discernibility brought out something with regard to force itself: it has the character of being a cause and is related to movement, namely, movement in another. Aristotle saw this in a decisive essential moment and brought it to conceptualization. It is necessary to hold fast to this result and for once to deal with it thoroughly. If the self-evident is to become questionable, then it is necessary (1) not to digress prematurely in interpreting the Aristotelian treatise, as is the popular fashion, by finding assurance in a certain similarity with the principles and assertions of later thinking, thus taking the question as settled; (2) but to sharpen our vision for something remarkable, for what confronts us in the Aristotelian treatment of δύναμις.

What we have to take note of here is the very way in which Aristotle develops this essential delimitation of δύναμις, making it more or less plastic. To put it defensively: it is not because he defines more thoroughly the individual elements of the determination (ἀρχή, μεταβολή, ἕτερον), but rather because he makes this essential delimitation visible as the guiding meaning for other corresponding meanings. And so in the same simple and demonstrative fashion other δυνάμεις are brought forth, but in such a way that their reference back to the first guiding meaning at the same time becomes visible; the guiding meaning is then for its part determined more clearly.

Before we pursue the further progression of the determination of δύναμις, let us once more point out the ambiguity of the Aristotelian formulation of the guiding meaning. Δύναμις is ἀρχὴ μεταβολῆς ἐν ἄλλῳ ἢ ᾗ ἄλλο. It is tempting to understand and translate this as follows: the origin for a change in another; an example of such an ἀρχή would be the potter at his wheel—the from-out-of-which, that from which change ensues: ἐν ἄλλῳ, namely, in the unformed lump of clay. This lump of clay is the other, ἄλλο, in which the change into the formed product, the jug, occurs. This now implies, however, that that from out of which the change ensues is likewise another: the potter.

And yet we could also understand the definition in this way: the origin of a change, which origin is in another; the ἐν ἄλλῳ as related to ἀρχή. And then μεταβολή must not be taken as change in the sense of a mere forward-moving modulation, of an alteration, as in the saying "The weather has changed." Rather, it must be taken in the meaning which it primarily has in Greek: to transpose or to shift, for example, to shift the sail, to transpose goods, thus in an "active" sense. I have decided upon the latter interpretation: δύναμις is that on behalf of which a transposition ensues, and in such a way that this from-out-of-which (ἀρχή) is in another being (ἐν ἄλλῳ) than what is transposed. See as well the example at the beginning of Δ 12; there it is very clear: the art of building (οἰκοδομική) is a δύναμις, ἣ οὐκ ὑπάρχει ἐν τῷ οἰκοδομουμένῳ—which is not present in what is built but rather again ἐν ἄλλῳ. But then again, if that from out of which the transposition ensues is the same as that in which the change ensues, then the ἀρχή is ἀρχή only to the extent that the same being is taken in a different respect (ᾗ ἄλλο). The human is the origin of a medical treatment not to the extent that it is a sick human but rather to the extent that it, as a human, is a doctor. By virtue of this difference of point of view within the same being, a one and an other is attained.

There appears at first to be essentially no difference in what results if the ἐν ἄλλῳ is referred to μεταβολή. And yet it belongs primarily to ἀρχή: (1) with regard to the meaning of μεταβολή, (2) according to the explicit formulation at the beginning of Δ 12; and (3) this interpretation is demanded by the concrete problematic of δύναμις τοῦ ποιεῖν, as we shall soon see.

§ 10. The ways of force

a) Bearance and (prior) resistance. Effect as the being of the things of nature (Leibniz)

After advancing the guiding meaning of δύναμις, Aristotle continues Θ 1, 1046a11-16:

ἡ μὲν γὰρ τοῦ παθεῖν ἐστὶ δύναμις, ἡ ἐν αὐτῷ τῷ πάσχοντι ἀρχὴ μεταβολῆς παθητικῆς ὑπ' ἄλλου ἢ ᾗ ἄλλο· ἡ δὲ ἕξις ἀπαθείας τῆς ἐπὶ τὸ χεῖρον καὶ φθορᾶς τῆς ὑπ' ἄλλου ἢ ᾗ ἄλλο ὑπ'ἀρχῆς μεταβλητικῆς. ἐν γὰρ τούτοις ἔνεστι πᾶσι τοῖς ὅροις ὁ τῆς πρώτης δυνάμεως λόγος.

"The one (way of being a force) is namely a force of tolerating, that which itself is in the tolerant as the origin of a tolerable change, tolerable from another, or else from itself, to the extent that it is another. The other (way of being a force), however, is the behavior (the composure) of intolerance against change for the worse and against change in the sense of annihilation by another, or to the extent that what undergoes change is another, by the other precisely as the origin of a possible change. In all these delimitations (of the ways of being a force) there is at bottom what we addressed as the initial, guiding meaning of δύναμις." (See also Θ 12, 1020a4: ὁ κύριος ὅρος—the ruling, dominating delimitation.)

We have thus two δυνάμεις: (1) ἡ μὲν δύναμις τοῦ παθεῖν—the power to tolerate something from another; (2) ἡ δὲ ἕξις ἀπαθείας τῆς ἐπὶ τὸ χεῖρον—the behavior of intolerance (insufferability) in reference to a change for the worse or even in the sense of annihilation by another.

With regard to (1): the power to tolerate. We may not yet here think of tolerating in the sense of painful enduring. Tolerating, not as suffering and not at all as hurt, but rather tolerating has here the sense of "he can't stand it": he does not readily allow it. Tolerating in the sense of allowing. The lump of clay tolerates something; it allows the formation, that is, it is malleable as a way of force. In this allowing, the lump of clay itself participates in its own positive way. It tolerates the forming because, as it were, it can tolerate it, because from out of itself it has a certain sym-pathy for this. This tolerating is a bearing, not in the sense of "bearing fruit," not in the sense of "bearing as giving forth," but rather bearing in the sense of allowing. We can capture this way of δύναμις with the word: *bearance* [*Ertragsamkeit*]. A particular kind of bearance is an enduring, endurance. The allowing of the forming is, however, at the same time a not-being-against in the sense of: doing nothing against . . . , simply letting it happen on its own; a δύναμις in the sense of a μὴ δύνασθαι (see Θ 12, 1019a28-30). This kind of force can be seen in that which somehow sustains damage through contact with it, in something which is shattered (κλᾶται) or pulverized (συντρίβεται) or twisted (κάμπτεται). Here tolerating, as not enduring, not sticking it out, is a being damaged, a loss. In contrast, with the transformation of the clay into the bowl, the lump also loses its form, but fundamentally it loses its formlessness; it gives up a lack, and hence the tolerating here is at once a positive contribution to the development of something higher.

What the δύναμις τοῦ παθεῖν has, in the sense of bearance, is παθητικός—that which in general can run up against something, that which can "undergo" something. This is accordingly a non-resisting, something in which resistance remains absent; τὸ μὴ δύνασθαι καὶ ἐλλείπειν τινός (ibid.). This is fragility in the broad sense of the word, not just breakability, but the non-durable. It is to be noted that this account of παθεῖν runs up against a negative feature; the very opposite is the case with the kind of force that Aristotle introduces in connection with it, the ἕξις ἀπαθείας.

With regard to (2): this non-tolerating, non-enduring is resisting,

§ *10. The ways of force* 75

the self asserting, enduring, and coming through against damage and degradation, against even annihilation, thus *resistance* in general.

Bearance (formability and fragility) and resistance, both are types of δύναμις. Aristotle wants to show: ἐν γὰρ τούτοις ἔνεστι πᾶσι τοῖς ὅροις ὁ τῆς πρώτης δυνάμεως λόγος (Θ 1, 1046a15f.). We ask ourselves: according to this, what is the basic meaning of πρώτη δύναμις which must somehow already lie in these two? Up until now it has been stated as ἀρχὴ μεταβολῆς—the from-out-of-which for change; now it is that from out of which change is allowed, or else that from out of which such change is resisted. Here also there is a relationship to change, and in such a way that the reference mentioned first (ἀρχὴ μεταβολῆς) is already set down with it. How? That comes to light precisely by going back to the essence of δύναμις which has now become visible. For that which is the origin for a resisting, what resists, is in itself indissociably—not incidentally—*referred to* something which runs up against it, to such a thing which does something to it, which wants and ought to do something to it (ποιεῖν). In the same way: the fragile, that which does not hold up, decays, and is thereby "exposed"—to another which works on it. The δύναμις τοῦ παθεῖν has a reference to a δύναμις τοῦ ποιεῖν that inheres in its very constitution, a reference to doing. Thus it becomes clear: In the guiding meaning—ἀρχὴ μεταβολῆς—the being an origin is what it is for a ποιεῖν, for a doing; this means: μεταβολή must be understood in an active sense; ἀρχὴ τοῦ ποιεῖν or else τοῦ ποιουμένου. Ἀρχὴ μεταβολῆς means then: being an origin for a transposing pro-ducing, a bringing something forth, bringing something about. This means being an origin for having been produced, having been brought about. Accordingly, δύναμις τοῦ ποιεῖν καὶ πάσχειν is also discussed directly in the subsequent passage (lines 19-20).

But now in advancing the guiding meaning of δύναμις, Aristotle does not speak of δύναμις τοῦ ποιεῖν (or ἀρχὴ ποιήσεως). This is neither neglect nor lack of rigor. The presumed indefiniteness of the formulation ἀρχὴ μεταβολῆς instead bears witness to precisely the clairvoyance and the conceptual acuity of Aristotelian thinking. The omission of the determination ποιεῖν does not imply that ποιεῖν is thus excluded. Yet if it were present, this would only prompt the misunderstanding that it is valid simply to say: there are first forces of doing, and secondly also forces of suffering, and then, thirdly, of resisting. The characterization, "being the origin of a change," even in its apparent indefiniteness is not intended to sever prematurely the inner relation of δύναμις τοῦ παθεῖν to δύναμις τοῦ ποιεῖν and vice versa; it is intended, rather, to make a space available for it.

The entire connection is accordingly to be grasped in this fashion: the guiding meaning does not imply a δύναμις isolated unto itself (ποιεῖν), in addition to which then other further meanings are listed. Instead, the others in their very constitution refer to the first, and in such a way that precisely this reference also gives back to the guiding meaning its very sense and content. The πρώτη δύναμις is the basic outline of this essence into which the full content is then to be drawn; only this extraordinary sketch puts into relief the whole essence of δύναμις and thus the full content of the guiding meaning as well.

Thus one may not say that δύναμις τοῦ ποιεῖν is the first and decisive δύναμις, as if the others were formed after it, but just as decisive is the relation of δύναμις τοῦ ποιεῖν, the force of doing, of producing, to the δύναμις τοῦ παθεῖν, the force of bearing and resisting. And just this relation is not to be made clear in terms of δύναμις τοῦ ποιεῖν but instead in terms of δύναμις τοῦ παθεῖν. This one is mentioned first, first in the sense of the actual characterization.

And this is not accidental. Yet Aristotle gives no precise reason for it. He simply follows the natural constraints of the matter at hand. How so? Δυνάμεις as bearance, endurance, formability, and fragility, in themselves bear witness to the character of resistance and non-resistance. *That which resists* is the first and most familiar form in which we experience a force. This means we do not first experience resistance as such, but rather something which resists, and through this resisting, we then experience that from out of which a change ensues, or else does not ensue, namely a force.

From this one would like to surmise that in the end there is indeed some legitimacy in what we earlier rejected: that we do experience forces primarily in reference only to the subject—to our own doing— and in such a way that we experience forces from effects. Except now

§ 10. The ways of force

we stand before a completely different state of affairs: when we experience effecting forces through the experience of what resists, then resistance is not taken and explained as the effect of a cause that is hidden behind it, but rather what resists is itself the forceful and the force.

Furthermore, if we ourselves participate as subjects in this direct experience of forces, then the objective view of our own experience of this exhibits precisely this one thing: There is nothing here which has to do with a transfer of subjectively experienced forces onto the affected things, but just the reverse. In what resists, the forces press themselves upon us. We experience the forceful not first in the subject but in the resisting object. And in its resistance, we experience in turn its non-ability, its restriction. And only in this do we experience a wanting to be able, a tending to be able, and an ought to be able. This, of course, is not to be understood as though the concept of force is now conversely to be transferred from the objects outside onto the internal subject.

But even in this way the matter is still presented in a misleading way. One could say: Given that the forces are not projected and transferred as subjective experiences onto things, and given that we come across forces directly in what resists, then indeed one thing still does hold, that we do transfer the forces that are experienced, that is to say, we do transfer the effect-relation that thingly forces have for us, the effect-relation of objects for subjects, onto the relations that objects have among themselves. To give an example: We speak therefore of forces in nature and of the interactive effective connections of forces among themselves—such as thrust and counterthrust—only because we wrest, as it were, the recoiling relation that things have for us both into and out of the relationship which holds between things themselves. This means: we do not experience directly among the things which surround us and in the relations which hold between them such a thing as the relation of resistance and the effects which run counter to it. And yet we do experience this. But we have to learn to see it if we are to achieve the level at which the Aristotelian considerations are working themselves out.

Thus what is at issue here is perceiving and comprehending forces and counterforces in nature itself, among the things of nature as such, and not just to the extent that they are objects for us, that is, not just to the extent that they are resistant. With this we do not at all want to get involved with the pre-scientific and extra-scientific experience of nature, with what makes itself manifest in a so-called nostalgia for nature. Aside from the difficulty of actually comprehending and grasping this philosophically, as a way in which nature becomes manifest, we set these thoughts aside because these kinds of experiences are especially subject to a current suspicion. One believes that basically this has to do simply with the empathy that subjective experiences and moods have for things, which in themselves and in truth present only the colliding and the displacement of ultimate particles of matter.

*

The guiding meaning of δύναμις is ἀρχὴ μεταβολῆς ἐν ἄλλῳ ἢ ᾗ ἄλλο. We attempted to show to what the ἐν ἄλλῳ belonged, namely to ἀρχή. For the understanding of this guiding meaning and for the consequences of what Aristotle achieved, we must take note of and follow up on two things: (1) the completely preliminary point of departure which is not burdened by far-reaching theories; (2) the realization of the whole delimitation (ὁρισμός). In any case, a dissolution into elements is not what is at issue here. The procedure does not permit that it be characterized in a positive manner with one word; it progressively makes a basic outline more distinct. The actual consideration begins with the differentiation between two types of δύναμις: (1) ἡ μὲν τοῦ παθεῖν—bearance, endurance (or else fragility); (2) ἡ δὲ ἕξις ἀπαθείας—resistance. Both are modes of the from-out-of-which for a change, that from out of which such a change is determined. It is now said of these two as a decisive characterization: ἔνεστι ὁ τῆς πρώτης δυνάμεως λόγος. How so? It is striking that Aristotle begins with a discussion of δύναμις τοῦ παθεῖν, with πάσχειν, suffering. The common opposing concept is ποιεῖν, doing. And yet this is not at all what is under discussion, nor does it come to be later. And when this is discussed, it is done as if this kind of δύναμις had already been introduced. And so it has been, if it actively resonates in the guiding meaning. But if that is so, then the ἐνεῖναι of this ποιεῖν is not to be thought in the παθεῖν that is now actually being discussed, as if it were the genus. Doing is in no way the genus of suffering, as supposedly one kind of doing. Instead the reference of δύναμις τοῦ παθεῖν to πρώτη δύναμις is a different and more essential one: Both what resists and what endures are in themselves related to that which occupies itself with them. The guiding meaning is not such that it is neither ποιεῖν (doing) nor παθεῖν (bearing); it does indeed express the δύναμις τοῦ ποιεῖν, but in such a way that it remains open for the inclusion of the essential reference back to itself of δύναμις τοῦ παθεῖν. This is also the tendency of our subsequent discussion; as we proceed through the different ways of δύναμις, we do not leave the guiding meaning behind but continually return to it. —In the situation in which we experience force, what resists shows itself to have a certain priority. We now want to see how we also experience directly in the oppositional relation between things themselves something like resistance and confrontation, thrust and counterthrust.

Hence we now want to survey one realm of our comprehension of nature which is not so readily exposed to the suspicion that it is merely a subjective mood of empathy. I mean the field of mathematico-physical research into nature. We take a question from this realm which historically and essentially stands in the most intimate relationship with the Aristotelian distinction between δύναμις and ἐνέργεια. This is the *introduction of the concept of force into mathematical physics by Leibniz*, a step which was taken in continual confrontation with Descartes and Cartesianism and which in its most apparent aim, at any rate, strives toward an adequate determination of the dimension of the possible measurement of forces in the context of the movement of material nature. This is a step which of course does not obtain its guiding motives from questions of physics, but from the basic question of philosophy concerning the essential determination of being in general and as such. From out of the multilayered entirety of the relevant questions, we shall bring out only one concern. It pertains to the fundamental question concerning the characterization of the being of natural beings, that is, of the things of nature which are for themselves present and are there in advance, namely the being of "substances."

80   *Metaphysics* θ *1*.

Descartes asserts that what is distinctive in the *res naturae* is *extensio*, extension; the natural thing is *res extensa*. Spatial expansion is indisputably one characteristic belonging to the things of nature experienced by us. But why did Descartes make this so distinctive, putting it forth as the fundamental determination? His intention here is decisively a critical one, simultaneously negative and positive; negative: against the explanation of nature in medieval scholasticism, against the assumption of concealed forces; positive: with the intent of thus achieving a determination of the things and processes of nature, their movement, that makes scientific knowledge possible, with its corresponding provability and determinacy. Scientific knowledge is, however, mathematical. And so Descartes asks: How must nature be posited such that it can be recognized scientifically and mathematically?

From the very beginning this is guided by the idea of pure certainty, of an absolute indubitable knowledge. That which is able to be at all must be adequate to this idea of knowledge, namely to the idea of being known, which is indissociable from it. This idea of knowledge and certainty was developed by Descartes in his *Regulae ad directionem ingenii*. What is fundamentally given is *intuitus* (intuition), that is, an *experientia* (experience) which has the character of a *praesens evidentia* (of direct evidence). This is the construction of an idea of knowledge which presents itself first of all in the mathematical. But because mathematical knowledge is primarily related to what is spatial, extension is put forth as the primordial characteristic of substance.

Scientific knowledge of nature comprises that which necessarily and generally must be asserted of natural things. As what is asserted, this is the truth concerning nature, which means what and how nature is in truth. The question of a possible scientific knowledge of nature is in this formulation at once the question concerning the true being of the beings which we call nature.

Leibniz turns against this determination of the being of natural things and says: The being of these substances does not lie in extension (*extensio*) but in activity [*Wirken*] (*actio, agere*). Two things must be noted in this new articulation of the being of natural things: (1) With this Leibniz does not want to eliminate the determination of *extensio*.

§ 10. The ways of force

This remains intact, but in such a way that it is acknowledged as grounded upon a more original determination of being in the sense of acting. (2) This concept of acting is now grasped in the context of our present problem such that the beings which are determined in this way now more than ever admit a mathematical determinacy. In this way it comes about that, in comparison with Descartes, a much more intimate and essential connection becomes possible between the mathematical method of measuring movement (infinitesimal calculus) and the kind of being which is knowable, something which we shall not enter into here.

Leibniz presented the considerations which are relevant here on different occasions and repeatedly in confrontation with Cartesianism. This is clearest in his *Specimen dynamicum pro admirandis naturae legibus circa corporum vires et mutuas actiones detegendis et ad suas causas revocandis*, written in 1695.[1] In this *Specimen from the doctrine of dynamics* Leibniz says clearly and concisely: "*cuando agere est character substantiarum, extensioque nil aliud quam jam praesuppositae nitentis renitentisque id est resistentis substantiae continuationem sive diffusionem dicit, tantum abest, ut ipsammet substantiam facere possit*" (loc. cit., p. 235). "For activity is the originally demarcated and thus the distinguishing essence (character) of substances; extension, however, means nothing other than the continual repetition or expansion of a substance which has already been posited in advance as striving and striving against, that is, as resisting; extension is thus a far cry from being able to constitute the essence of substance itself." The *agens* is *nitens* and *renitens* (*resistens*), the *actio* a *nisus* and *conatus*, an inclination and a striving. *Hic nisus passim sensibus occurrit, et meo judicio ubique in materia ratione intelligitur, etiam ubi sensui non patet* (p. 235). This characteristic of the being of substance—striving, striving against, resisting—"often directly confronts experience (we come across such a being), and even when this characteristic does not manifest itself to sensible experience, it is, I suspect, through deliberation seen everywhere as that which belongs to matter." Such a thing as

---

1. *Leibnizens Mathematische Schriften*, ed. C. I. Gerhardt (Berlin and Halle, 1849-63), vol. VI, pp. 234-46 (part I).

resistance among things themselves and against themselves is able to be experienced, or rather, to be made visible as what necessarily belongs to things. But resistance is not a simple indifference to movement, not a mere "non-participating," but a striving against, and this means something proper, from out of which something determines itself in another thing. If we were to assume that the essence of substances, and along with this their movement, are to be determined principally and exclusively through *extensio*, if, therefore, there were no resistance (*renisus*), then it would be incomprehensible why a large body, for example, should not be dragged along by the movement of a far smaller one running into it, without the smaller one thereby suffering any deceleration. The colliding body would have to achieve this effect fully (cf. p. 241). In other words, in perceiving a collision, we experience in advance, in what lies in the being of these bodies, more than a mere alteration of their relationships of extension and location, that is, more than what the Cartesian theory of *extensio* would like to admit. We experience one body being stopped by another, we experience the colliding into . . . . Belonging to the being of a body is a *vis*, and here there is a particular *vis passiva*—the *derivativa* in distinction to *primitiva* (cf. p. 236f.).

Leibniz was fully aware of the basic implications of this new formulation of the being of substances—of all substances, that is, all beings, and not just the material things of nature. He was at the same time resolved to bring the previously misunderstood Aristotelian doctrine of δύναμις and ἐνέργεια into its own right and to restore its true meaning. *Et quemadmodum Democriti corpuscula, et Platonis ideas, et Stoicorum in optimo rerum nexu tranquillitatem nostra aetas a contemtu absolvit, ita nunc Peripateticorum tradita de Formis sive Entelechiis (quae merito aenigmatica visa sunt vixque ipsis Autoribus recte percepta) ad notiones intelligibiles revocabuntur* (p. 235). "As it is with the doctrine of Democritus regarding small particles and with the doctrine of Plato's ideas, and with the doctrine of the Stoics regarding the tranquility of the mind, which arises from insight into the best order of things, as it is with all these doctrines, which in our age have been absolved from contempt, now, too, is the traditional doctrine of the Peripatetics regarding the forms and the entelechies being made here conceptually intelligible, a doctrine which rightfully has been considered enigmatic but which until now has been barely comprehended, even by the authors themselves."

For us what is important here is to see how the renewal of the Aristotelian doctrine of δύναμις and ἐνέργεια appeals to definite aspects of the constitution of beings themselves. To this belongs the possibility of experiencing the collision and the recoil of one body upon another, and so of the resistance of things among and against themselves. Here there is at first only the one proviso: not to prematurely explain away or even downright ignore the simple facts of experience for the sake of some theory.

If we were to say that resistance is able to be experienced not only from the relatedness of things to us, not only within the realm of our ability to be confronted by things, but also among the things themselves in their oppositional relations, then it would still be undecided whether this manifestness of the striving and counterstriving of things among themselves does not stand under the very determinate conditions of our Dasein. But if this is the case, then we are still a long way from saying that the concept of force arises from the subjective experiences of one's own ability. And there is the further question of whether the objectivity of what is experienced is even addressed by referring the possibility of experiencing the thrust and counterthrust of things among themselves to certain conditions.[2]

However these questions are to be posed and developed, the precondition for their treatment is always that we first of all become clear-sighted again with regard to the phenomena which we meet along the way. And to these belong especially the two modes of δύναμις with which we have become acquainted (Θ 1, 1046a11ff.): the force of bearing (δύναμις τοῦ παθεῖν) and of resisting (ἕξις ἀπαθείας). We saw, moreover, how the most proper conception of these δυνάμεις carries with itself and in itself the reference to the δύναμις which Aristotle adduces first of all. For Aristotle, too, designating a further way of δύναμις κατὰ κίνησιν turns

---

2. Being-obstructed—in alignment with being-unobstructed: it can be dealt with . . . ; the for the sake of, care. Concerning the experience of resistance, see *Sein und Zeit* I (Halle a.d. S., 1927), § 43, p. 200ff.

b) The how which belongs to force

1046a16-19:

> πάλιν δ' αὗται λέγονται δυνάμεις ἢ τοῦ μόνον ποιῆσαι ἢ [τοῦ] παθειν ἢ τοῦ καλῶς, ὥστε καὶ ἐν τοῖς τούτων λόγοις ἐνυπάρχουσί πως οἱ τῶν προτέρων δυνάμεων λόγοι.

"Once again the forces (δυνάμεις) are now named which have already been introduced and as such are understood either just simply in reference to that for which forces are, for doing or for suffering, or else in reference to what is 'in the right way,' so that even in understanding this meaning of δύναμις the meanings of the aforementioned are also in a certain way understood along with it."

The πάλιν takes up the heretofore discussed meanings in order to put into relief once again a characteristic rooted in the guiding meaning but which at the same time determines this guiding meaning from a new perspective, thereby bringing it to a fuller determination. We are told what this means in Δ 12 by means of an example (1019a24ff.): ἐνίοτε γὰρ τοὺς μόνον ἂν πορευθέντας ἢ εἰπόντας, μὴ καλῶς δὲ ἢ μὴ ὡς προείλοντο, οὔ φαμεν δύνασθαι λέγειν ἢ βαδίζειν. "For at times we say of those who do walk or speak but not in the right way, or else not in the way in which they intended, οὔ δύνασθαι—that they cannot walk or speak." Thus we say of a poor speaker: He cannot speak. And yet, if we consider the matter simply and directly, he certainly does speak, perhaps even too much. But we say likewise of a runner who in fact zealously moves his legs and even goes somewhere that he nevertheless cannot run; he is lacking the right way, and lacking stamina; he is not accomplished in his own affair—and so it is in each case different. Having the power for something . . . means here: being accomplished in something, succeeding in the right way. Ability in this sense is mastering, being the master over . . . , mastery. We simply say of a good speaker: He is a speaker. Being means here: having in the right way the power to do the task at hand. Having the

§ 10. The ways of force

power for something is properly a force first when it is in the right way.

To having the power for something, there belongs necessarily a *how* which can modify itself in such and such a way but which carries within itself the *claim to a possible fulfillment*—καλῶς ἐπιτελεῖν. Precisely in the concept of δύναμις κατὰ κίνησιν there is then also a reference to τέλος which inheres in its very constitution. This does not mean anything like "purposeful behavior," but rather: an inner ordering of something toward an end, a conclusion, an accomplishment. Hence δύναμις implies the moment of being on the way toward something, of the oriented striving, oriented toward an end and a being accomplished—and hence there belongs to the inner structure of δύναμις the character of "in such and such a way," "this way or that," in short: the how. This now holds also in the same way for the δύναμις τοῦ ποιεῖν as well as for the δύναμις τοῦ παθεῖν and δύναμις ἕξις ἀπαθείας.

The meaning of δύναμις which is introduced under πάλιν is not a new kind which stands over and against the others, but is to be brought back to the guiding phenomenon and to be conceived as belonging to it in its very constitution. With this the insight into the full content of δύναμις as ἀρχὴ μεταβολῆς becomes more focused. Being as ἀρχή, being the origin for . . . , does not mean, then, a thing or a property from which something proceeds. Instead, being an origin for something other is in itself a proceeding to the other. This proceeding toward can succeed or fail in attaining what it goes after; it must always already prepare itself for a particular mode and manner of itself. The καλῶς does not signify an addition but rather points to a characteristic which, although variable in itself, belongs to the essence of δύναμις. Hence, the power for something is always a not falling short of a definite how. In the essence of force there is, as it were, the demand upon itself to surpass itself.

Hence if we speak simply of one δύναμις, of a μόνον δύνασθαι, then we are turning away from the characteristic of καλῶς. More exactly: We are leaving the essential and indissociable how indefinite and indifferent; we are not keeping the full essential content of δύναμις in view.

*

An attempt was made to indicate concisely just how the assignment of force as a basic characteristic of even the material things of nature appeals to the basic experiential relationships of thrust and counterthrust among the things themselves. But Leibniz not only defended this formulation of *substantia* as *actio* against Descartes's definition of *extensio*, he also grounded it at the same time fundamentally and philosophically through the doctrine known as "monadology." One usually sees the import of this doctrine—and this holds for philosophical reports as well—in Leibniz's assumption that all things are, so to speak, besouled: everything that is, is endowed with force—forces are then a kind of parasite and bacteria that have worked their way in everywhere. In truth, however, according to the principle of monadology, it is not the individual beings which are endowed with force, but rather the reverse: force is the being which first lets an individual being as such be, so that it might be endowed with something at all. Force is thereby comprehended as tendency, as *nisus* (*nitens* and *renitens*). In principle, then, the imputation that natural things are determined by forces is only an essential consequence of the determination of substance as monad, *vis*.—Leibniz appeals explicitly to the connection his doctrine has to that of Aristotle, which is to be brought to its own efficacy. It would be a misunderstanding if we wanted to lecture to Leibniz that his conception is incorrect and unhistorical, if we wanted to take issue with him that he in fact interprets only his own monadological doctrine of being back into the philosophy of Aristotle. The Leibnizian comprehension of Aristotle is admittedly not correct, but by no means in the sense that it is incorrect, for "correct and incorrect" is no standard at all when it comes to true historical knowledge. But to be sure, the Leibnizian comprehension of Aristotle is an essential one—and thereby a genuinely historical one. It is a conception that has something which can encounter the past, something to which then this past alone answers and expresses itself. What we gather from the Leibnizian appeal to the ability to experience thrust and counterthrust is this: The regress to the "subject" is not required in order to comprehend the basic outline of the essence of force. —The phenomena of force, bearance (δύναμις τοῦ παθεῖν) and resistance (ἕξις ἀπαθείας), are for this reason first mentioned by Aristotle in order to make visible the inner reference to the force of doing and forming (δύναμις τοῦ ποιεῖν), and so in order to interpret δύναμις as ἀρχὴ μεταβολῆς in an active sense. The καλῶς ἐπιτελεῖν also turns upon this relating back. But the inner grounding of the necessary belonging of the how to force in general has not yet been explicitly discussed (see below, p. 134f.).

§ 11. The unity of the force of doing and bearing: the ontological and the ontic concepts of force and their inner adhesion

What now follows in the text (Θ 1, 1046a19-29) at first looks like a repetitive summary.

> φανερὸν οὖν ὅτι ἔστι μὲν ὡς μία δύναμις τοῦ ποιεῖν καὶ πάσχειν (δυνατὸν γάρ ἐστι καὶ τῷ ἔχειν αὐτὸ δύναμιν τοῦ παθεῖν καὶ τῷ ἄλλο ὑπ' αὐτοῦ), ἔστι δ' ὡς ἄλλη. ἡ μὲν γὰρ ἐν τῷ πάσχοντι· διὰ γὰρ τὸ ἔχειν τινὰ ἀρχήν, καὶ εἶναι καὶ τὴν ὕλην ἀρχήν τινα, πάσχει τὸ πάσχον καὶ ἄλλο ὑπ' ἄλλου· τὸ λιπαρὸν μὲν γὰρ καυστόν, τὸ δ' ὑπεῖκον ὡδὶ θλαστόν· ὁμοίως δὲ καὶ ἐπὶ τῶν ἄλλων. ἡ δ' ἐν τῷ ποιοῦντι, οἷον τὸ θερμὸν καὶ ἡ οἰκοδομική, ἡ μὲν ἐν τῷ θερμαντικῷ, ἡ δ' ἐν τῷ οἰκοδομικῷ. διὸ ᾗ συμπέφυκεν, οὐδὲν πάσχει αὐτὸ ὑφ' ἑαυτοῦ· ἓν γὰρ καὶ οὐκ ἄλλο.

"Evident now is the following: A force for doing and for tolerating is at one time (in one respect) such that one single force (as one and the same) is both, . . . but at another time it is such that each of the two is another (such that it is both, in each case as other in another)." Now the explanation for the first possibility (δύναμις ὡς μία): "Having the force for something (for bearing) means, namely, both: having the force to bear, as well as: having this force by virtue of the fact that another suffers from that which has the force to do something." Now comes, with ἡ μὲν (a22) and ἡ δέ (a26), the explanation of the second possibility (δύναμις ὡς ἄλλη): "The one δύναμις is in that which bears for the reason that this has something like an origin, a beginning, for something else, and for the reason that the material, too, is something like an origin, a beginning for something; and thus for the reason that the one that bears bears something, and this as the one by virtue of the other. The oily is namely burnable, and the yielding, whatever it may be, is breakable; and so it is in the same way with other things. The other δύναμις, however, is present in what is producing, for example, the warm in what is warming, the art of building in the one who knows how to build. —Because of this, to the extent that what is producing (doing) and what is bearing (suffering) are present together as one and the same being, this being itself tolerates nothing by itself; for it is (then) one thing and not that one thing toward another."

What does Aristotle want to say with these sentences, which he adds to his characterization of the differing δυνάμεις? At first glance it is, in fact, only a summarizing review through which we learn that δύναμις as such is a δύναμις of doing, of production, and a δύναμις of suffering, of bearing; and the doing must always be in another being than what suffers, or else, where they are the same, as in the case with the doctor who treats him- or herself, there indeed both are the same human, the one who is treating, the doctor, and the one being treated, the patient, but in different respects—first the human as a person who practices a profession, and then as an occurring life exposed to peril.

It is, of course, in principle not precluded that such a summary be given; we find this quite often in Book Δ (cf. Δ 12). But the paragraph under discussion has such a striking and conscious arrangement that we may not evade the challenge to put the expressed content correspondingly into sharp relief. Δύναμις τοῦ ποιεῖν καὶ πάσχειν is being discussed. Heretofore this has not been discussed at all in this form. But this phrase already carries with it the problem that Aristotle wants to resolve in what follows.

For the question here is: When one speaks of the δύναμις of doing and of tolerating, are two δυνάμεις meant, two modes of δύναμις, or only one? If only one, then in what sense is δύναμις understood? If two, then how is δύναμις to be grasped in its unity?

Aristotle says simply: (δύναμις τοῦ ποιεῖν καὶ πάσχειν) ἔστι μὲν ὡς μία . . . ἔστι δ' ὡς ἄλλη (1046a19, a21/22). It is thus explicitly

§ 11. Force of doing and bearing

brought into tension with an ambiguity in the essence and the concept of δύναμις. The δύναμις τοῦ ποιεῖν καὶ πάσχειν is in a certain way *one*. This means: force is in itself force for doing, for producing; in the orientation toward what is to be produced, there is the reference to what can be produced. The producible is such an out of something. Being out of something is a mode of the from-out-of-which: as the pliant and formable, some kind of material, for example. The essence of force as force for producing also encompasses in a certain way this being "out of something" of producibility. The "out of something" of a thing's producibility is, however, δύναμις τοῦ πάσχειν. This δύναμις (as ἀρχή, as that from out of which something becomes producible) is implicated in the essence of the δύναμις τοῦ ποιεῖν. This is the case not only in general, but rather every particular producing, for example that of an axe, is related to stone, bronze, iron, and the like, but not to water, sand, or wood. Conversely, all these materials as potentially pliable implicitly refer to that which can be made of them through a definite way of producing, and so they refer to definite ways of producing and comporting with. Whether we know this explicitly in every case is irrelevant at this point.

The result of all this is that force in itself is the *relation* of the ἀρχή τοῦ ποιεῖν to an ἀρχὴ τοῦ πάσχειν, and vice versa. The essence of force in itself, in terms of its own essence and in relation to this essence, diverges into two forces in an originary way. This of course does not mean that a definite individual force directly at hand consists of two forces, but rather that this force in its essence, that is, being a force as such, is this relation of the ποιεῖν to a πάσχειν: *being a force is both as one*—ὡς μία. The relation [*Bezug*] is an *implication* [*Einbezug*]; force is implicating on the basis of its being directed outward and its going beyond. "Force" taken in this way, understood as being-force, is the "ontological" concept of force. "Ontological"—the traditional expression means the being of beings. Here it means: the force-being of every particular force, whether being in some way or other or possibly being. Force-being does not consist of two present-at-hand forces, but rather, to the extent that a force is present, there is in this being present the implicating outward directedness toward the corresponding opposing force. This is so because this outwardly directed implication belongs to the being-force of force.

If δύναμις, in contrast, is not understood ὡς μία but ὡς ἄλλη, if it is understood, that is, in such a way that with it *always one or the other* of the two forces belonging to force-being is intended as this individual force present for itself, then the expression δύναμις has, as we say, an ontic meaning. It does not mean force-being as being; it means, rather, a definite being, this being as the origin of a doing, or this being as the origin of a suffering. We then mean that respective being which has its share in force-being in a definite respect. We do not mean this force-being itself, but rather that which shares in it (the *subjectum*).

But Aristotle does not simply want to say that δύναμις in its usage has at times an ontic meaning, and at other times an ontological meaning. Instead it belongs to the essence of what we call force, that it must be understood in this ambiguity. If we experience or attend to a definite, present force, then we already understand being-force in advance, and to the extent that we understand this, we have in mind *along with it* that from out of which something is produced, something with which a beginning can be made, thus what is itself thereby in a certain sense a beginning, ἀρχή, namely ἀρχὴ τοῦ παθεῖν. Conversely, if we understand force-being in its essence, which means in the mutual implication of the opposing and outwardly directed origins, then we also already know that a being force [*eine seiende Kraft*] is *always* a kind of doing or of tolerating, and in such a way that what is doing is a different being from what suffers, or else when they are the same, the doer is the sufferer in a different respect. In this way from out of the essence of force it becomes clear why as ἀρχή it must necessarily be ἐν ἄλλῳ or ᾗ ἄλλο.

The sentence which closes the paragraph we have just dealt with (a27ff.: διὸ ᾗ συμπέφυκεν . . .) can be understood, if at all, only with regard for what has been presented, which means with regard for the fact that Aristotle wants to make known the *inner cohesiveness of the ontological and ontic concept of force*. The sentence is not at all comprehended in its content if one takes it, as interpreters do, only as a rewriting of the ᾗ ἄλλο. Aristotle wants to say: If force-being means the original unitary, implicating, and reciprocal relation of being an origin for doing and suffering, then this ontological unity of the reciprocal relation does not mean the ontic unity and convergence of ontic forces of mutually differing character. If one comprehends this unity in this way (ontically), then this "one" can so little provide the site for force-being, and for a relationship, that this "one" itself is not at all capable of tolerating something on its own behalf; and this also means that with reference to itself, neither is it a doer. The unity of force-being needs instead to be understood from out of the fact that this unity, as a unity of reflexive and inclusive relational being, demands precisely the ontic discreteness and difference of beings, which always persist with the character of force-being, that is, are the "subject" of force. Thus force does not consist of two forces, but rather, if force-being is in a being, then that being is split into two forces.

From this interpretation perhaps it has become clear why Aristotle does not simply define a sequence of individual "types" of forces, beginning with the force for doing, etc., as though his purpose would then be to seek the universal of these types. That he does not proceed in this fashion is grounded in his vigilant insight into the divisive, simple essence of force-being; that is, it is grounded in his successful entrance into the ontological and philosophical interpretation of essence.

§ 12. Force and unforce—the carrying along with of withdrawal. The full guiding meaning

Henceforth (in what the chapter has still to bring) another variation of the word δύναμις is explained in what appears again to be simply a loosely connected addition: force in the sense of unforce.

> 1046a29-35: καὶ ἡ ἀδυναμία καὶ τὸ ἀδύνατον ἡ τῇ τοιαύτῃ δυνάμει ἐναντία στέρησίς ἐστιν, ὥστε τοῦ αὐτοῦ καὶ κατὰ τὸ αὐτὸ πᾶσα δύναμις ἀδυναμία [not ἀδυναμίᾳ, H.]. ἡ δὲ στέρησις λέγεται πολλαχῶς· καὶ γὰρ τὸ μὴ ἔχον καὶ τὸ πεφυκὸς ἐὰν μὴ ἔχῃ, ἢ ὅλως ἢ ὅτε πέφυκεν, καὶ ἢ ὡδί, οἷον παντελῶς, ἢ κἂν ὁπωσοῦν. ἐπ'ἐνίων δ'ἐὰν πεφυκότα ἔχειν μὴ ἔχῃ βίᾳ, ἐστερῆσθαι ταῦτα λέγομεν.

"And unforce (forcelessness) and consequently also the 'forceless' is a withdrawal as what lies over and against δύναμις in the sense developed; hence every force, if it becomes unforce, that is, as unforce is in each case in relation to and in accordance with the same (with respect to that by which a force is a force, every force is unforce). Withdrawal, however, is stated and understood in multiple ways. Something is in a state of withdrawal (is affected by and is going through withdrawal) if it simply does not have something else; that is, when it does not have something even though it should have this something (what has withdrawn) according to its essence. (And again this not having is possible in various ways and respects:) either when what is affected by the withdrawal does not have what has withdrawn at all, or when it does not have this at the moment even though it could have this, or when it does not have this to some extent, for example when it does not at all have this, or finally when it does not have what has withdrawn in a certain way. We also say something is in a state of withdrawal if, through violence and on the basis of this suffered violence, it does not have what according to its nature it could and should have." (The parallel treatment in Δ 12, 1019b15-21.)

Here it is stated: In addition to force there is unforce, "*im-potentia,*" non-force. Yet this non- and this un- are not merely negations, but mean rather having withdrawn, "*being in a state of withdrawal*"—στέρησις. This, however, is understood in different ways corresponding to the possible various relations belonging to a relationship of withdrawal. This much is completely clear and understandable. If we do not go further, to be sure, we also remain outside the ontological and philosophical grasp of the matter which Aristotle wants to bring out here. In order to comprehend this, we shall begin with an explanation of στέρησις, but only to the extent that Aristotle himself introduces it in this passage. (Otherwise see Δ 22 and I.) We shall illustrate the essence of στέρησις by enlisting what Aristotle states in an example. In fact, we choose an example which is often cited and which at the same time is already related to the phenomenon it treats (δύναμις, force), namely the power of vision, sight, or else the lack of this power.

Something is in a state of withdrawal (ἐστερημένον) in different ways (Θ 1, loc. cit.); first, τὸ μὴ ἔχον—if it simply does not have something different from itself, for example, a stone and the power to see; ἐστερημένον is as well: τὸ πεφυκὸς ἐὰν μὴ ἔχῃ—when something does not have something different from itself although it should have this something according to its essence, should have what has withdrawn; for example, the human, for whom the power of vision, sight, is essentially appropriate, can lack the power to see. This not-having (of the human) mentioned second is now for its part still possible in different ways; first, ὅλως—the power to see is utterly withdrawn from the human when the human is blind from birth; then, ὅτε πέφυκεν—when the human does not have the power to see at the moment, although he or she could have it; for example, if a person, awake and looking around, cannot see for reasons of eye disease; καὶ ἢ ὡδί—or when the human does not have what has withdrawn to this or that extent, for example, a person with one eye; ἢ κἂν ὁπωσοῦν—when a person does not have what has withdrawn in a certain way, when he cannot see, for example, because it happens to be dark or because what is to be seen is otherwise covered.

We must again emphasize that by introducing examples such as this of the multiple ways of στέρησις, still nothing at all can be said concerning its essence. And even if the explicit treatment of στέρησις as such meets with only partial success in Aristotle and antiquity in general, still the movement which the discovery of this phenomenon brought to philosophy (up to Hegel) must not be forgotten. (The exactitude of the summarizing definition which follows is in any case secondary—and troubling over such a definition, even if successful, becomes disastrous if the full efficacy and consequence of this thought never actually comes to pass, or for some time has ceased to be.)

What relation, then, does στέρησις hold to our complex of questions concerning δύναμις? Does this occur merely to show that there is, in addition to force, unforce as well? No. Aristotle wants instead to say something else. This receives a concise formulation in the sentence at line 30f.: τοῦ αὐτοῦ καὶ κατὰ τὸ αὐτὸ πᾶσα δύναμις ἀδυναμία. "In relation to and in accordance with the same is every force unforce." And so the text is also clear; there is no need to improve it with the dative ἀδυναμίᾳ. What is emphasized here is the referring back of unforce upon the same thing by which force is force; what is emphasized is the constitutional belonging of unforce to the guiding meaning of force—as an inner variation of this, and even in differing respects that are already pre-given with the power to do something, each according to its particular content.

*

What appears at 1046a19-29 to be merely a redundancy is the first step to the unitary, essential determination of the phenomenon of δύναμις. The formulation δύναμις τοῦ ποιεῖν καὶ πάσχειν expresses an ambiguity. According to this, δύναμις means (1) ὡς μία—as something unitary: a being [*ein Sein*]; (2) ὡς ἄλλη—as always one and another: definite beings. With regard to (1), the force of doing and suffering as *one:* the one ἀρχή implicating the other—only with this implicating, reciprocal ἀρχή-character is the essence of δύναμις fulfilled. Regarding (2), if such a force comes to be, then it does so necessarily as two forces, or else as one being in two different ontic respects. Thus this passage does not argue for the ontic coexistence of two present forces as though they were one. Further, unforce is distinguished from force. The question here is once again whether only a modification is set forth, or whether this is a reflexive questioning for the comprehension of the entire essence. Στέρησις, the withdrawal of force, means (1) μὴ ἔχον—a simple not-having, (2) τὸ πεφυκὸς ἐὰν μὴ ἔχῃ—when something does not have what it should have, and this in differing ways and respects.

The decisive thesis reads (a30-31): "Every force is unforce with reference to and in accordance with the same thing." This states that unforce is nevertheless bound to the realm of force that remains withdrawn from it. That from which something has withdrawn is related in and through this withdrawal precisely to that which has withdrawn. And despite the negative character of the withdrawal, this withdrawing relation always produces its own positive characterization for that which is in the state of withdrawal commensurate with the way of the withdrawal (which itself is still different in relation to one and the same thing). Aristotle brings up at Δ 12 a significant example (1019b18f.): οὐ γὰρ ὁμοίως ἂν φαῖμεν ἀδύνατον εἶναι γεννᾶν παῖδα καὶ ἄνδρα καὶ εὐνοῦχον. "For we are not inclined to

§ 12. Force and unforce

call the boy, the man, and the eunuch powerless to procreate in the same sense." In this example the modifications of withdrawal and the modes of a positive being [*Sein*] which are given with them can be easily demonstrated, and thus they elucidate the essence of στέρησις as a modification of possessing and having. We, too, still use today the expression "impotent" (powerless) in a distinctive and emphatic sense with reference to the power of procreation. This points to a special bond between "force" and "life" (as a definite mode of εἶναι, of being), a bond with which we are acquainted from daily experience and common knowledge, without scrutinizing its inner essence and ground.

And then finally the modification of force to un-force, from possessing and having to a withdrawal, is a more essential one in the field of force than in other phenomena. Δύναμις *is in a preeminent sense exposed and bound to* στέρησις.

One is inclined in this context to point out that this modification into withdrawal occurs in many other realms as well. For example, rest is for movement the corresponding phenomenon in a state of withdrawal, and precisely this relationship is often and readily cited as a characteristic example of στέρησις. See also Aristotle himself, *Phys.* Δ 12, 221b12f.: οὐ γὰρ πᾶν τὸ ἀκίνητον ἠρεμεῖ, ἀλλὰ τὸ ἐστερημένον κινήσεως πεφυκὸς δὲ κινεῖσθαι. "For not everything unmoving is at rest; rather, we call the unmoving at rest only when it is unmoving through being robbed of movement, and in such a way that what has been robbed is suited, according to its inner essence, to being moved." (Compare what was said earlier about τὰ μαθηματικά: ἀκίνητα—the mathematical is unmoving inasmuch as it is outside the possibility of movement entirely, and only then is it also outside the possibility of rest.) Likewise is σκότος, darkness, στέρησις for φῶς, light (*De an.* B 7, 418b18f.), and keeping silent is a στέρησις of speech, silence a στέρησις of noise. Thus one does not see how force and unforce carry along with them the "steretic" relationship in a preeminent sense. Neither does Aristotle say anything about this. He certainly does not. And yet we still need to pursue this question further, not only because it is of far-reaching importance for the potential understanding of the essence of δύναμις κατὰ κίνησιν—and even of more importance for δύναμις ἐπὶ πλέον—but also because what is touched upon in this question leads directly to what Aristotle discusses subsequently in chapter two.

We saw that δύναμις in itself is at once δύναμις τοῦ ποιεῖν καὶ πάσχειν, that the relation to what can bear, to what endures, what resists, belongs to the essential structure of force. Not that it could ever be predetermined or directly discerned which beings withstand an effective force in their character of being able to bear; what is decisive is that what can bear belongs somehow to the realm of a force of producing. Every such force delineates a realm for itself, within which it dominates that for which it is, and what it is, namely force. Force then always dominates itself in a peculiar sense. Every force accordingly has a character of *possessing* that is difficult to grasp with sufficient generality; this character of possessing is precisely this implicating delineation of its realm. Hence a *losing*, and so a distinctive way of withdrawal, is in an emphatic sense capable of corresponding to this characteristic of possessing. The steretic alteration of force into unforce is accordingly of a different kind from, say, the turn from movement toward rest, not only because force and movement are different according to their particular content but because the proper possessive character of force is more inwardly bound up with loss and withdrawal.

The Aristotelian proposition, πᾶσα δύναμις ἀδυναμία (see p. 93), does not mean to say that wherever a force is at hand, there factically and necessarily an unforce is also at hand, but rather that every force is, if it becomes unforce, the loss of its possession. It is unforce by virtue of one and the same thing by which force is forceful. And, to repeat, it generally holds for every στέρησις that it pertains always to a δεκτικόν—what something can take to itself and keep; the δέχεσθαι, the taking-to-oneself, is in fact proper to δύναμις in a distinctive sense, so much so that δύνασθαι and δέχεσθαι at times become synonymous in the ancient Greek language.

Looking back, we now see, however, that naming ἀδυναμία and ἀδύνατον no more means a mere "enumeration" of some kind of force than did the earlier references to δύναμις τοῦ παθεῖν and ἕξις ἀπαθείας. Instead, it makes visible, as did these, a necessary structural element of the full essence.

§ 12. Force and unforce

If we were once again briefly to list *all the essential elements*, they would be: to force there belongs the from-out-of-which of a being-out-toward, and the reciprocal relation to what bears in the broadest sense. Each of the two exhibits the character of an implicating delineation which carries along with it as preeminently possessive the possibility of loss and withdrawal. The outwardly directed implication persists in this way in some kind of definite or non-differentiated how. The ἐν ἄλλῳ which belongs to ἀρχή as a determination does not simply designate the force that is articulated in this way as a thing that is present somewhere else, but means rather precisely this: Although the from-out-of-which is in another being, this other that is defined in this way is precisely that which has in itself a relational and governing reference, and so a wide-reaching range [*Reich-weite*]. With a full understanding of the preceding guiding meaning, we need to transfer all these attained characteristics of force back to and into this guiding meaning, in order to come up to the level of understanding that Aristotle demands when he says that the δυνάμεις κατὰ κίνησιν are addressed πρὸς τὸ αὐτὸ εἶδος—πρὸς πρώτην μίαν (cf. *Met.* Θ 1, 1046a9f.).

But then how do things stand with the determination of μεταβολή, which also appears in the guiding meaning (ἀρχὴ μεταβολῆς)? With the full explanation of the essence of force, it has, as it were, fallen by the wayside. And consequently neither have we learned anything about the extent to which the δύναμις under discussion is precisely δύναμις κατὰ κίνησιν. To be sure, this was not the topic. It is true that Aristotle does not deal with μεταβολή as such, but he does in fact deal with it inasmuch as δύναμις is defined as its ἀρχή. For what does it mean that δύναμις is the from-out-of-which, which implicates into its own realm that which in itself is able to bear? This indeed says only that force, on the basis of its essence, first provides *a possible site for a change* from something to something. To say that what can endure is exposed to something which works it over means: something like change is already and necessarily signified in this reciprocal relation, both what permits being formed into shape as well as the forming production. Hence μεταβολή in the fully understood guiding meaning no longer means one-sidedly only the active transforming; neither is it passive bearing simply appended on to this. Instead it means the reciprocal relation of both as such. In this sense μεταβολή is indeed implicitly discussed in the preceding: it has so little fallen by the wayside that pointing out these references—δύναμις τοῦ ποιεῖν καὶ πάσχειν ὡς μία—brings it into view.

Only now do we first correctly understand what is said in κατὰ κίνησιν, force understood in terms of movement. It does not simply mean, as it appeared initially, that what is at issue here is the force that can be discerned in a being that is found in movement. Instead, δύναμις κατὰ κίνησιν is of the sort whose essential structure is co-given in the basic phenomenon of μεταβολή, precisely in the reciprocal relation between δύναμις τοῦ ποιεῖν and τοῦ πάσχειν—understood in terms of such movement and with an eye toward it. It is no accident that Aristotle explicitly comes to speak of the relation between ποίησις and πάθησις (or ποίημα and πάθος) precisely where he expressly speaks of movement (cf. *Phys.* Γ 3).

While an enumeration and a grouping of the types of forces is not the intention of this first chapter which we have now discussed, the theme is rather precisely the unity of the articulated essential structure of force in general. Likewise, the following second chapter aims toward a division of δύναμις κατὰ κίνησιν.

Chapter Two

Metaphysics Θ 2. The Division of Δύναμις κατὰ Κίνησιν for the Purpose of Elucidating Its Essence

§ 13. Concerning λόγος (conversance) and soul. The divisions: "conversant/without conversance" and "besouled/soulless"

1046a36-b2: ἐπεὶ δ'αἱ μὲν ἐν τοῖς ἀψύχοις ἐνυπάρχουσιν ἀρχαὶ τοιαῦται, αἱ δ' ἐν τοῖς ἐμψύχοις καὶ ἐν ψυχῇ καὶ τῆς ψυχῆς ἐν τῷ λόγον ἔχοντι, δῆλον ὅτι καὶ τῶν δυνάμεων αἱ μὲν ἔσονται ἄλογοι αἱ δὲ μετὰ λόγου.

"Since, then, some (those forces understood with respect to movement) are present in beings without soul by way of belonging to them and co-constituting them, whereas others are present in the besouled, that is, in the soul as such, indeed, in the kind of soul that has in itself a discourse, it is thereby evident that some of the forces are also without a discourse, whereas others are with a discourse (directed by such a discourse)."

Thus, the chapter opens up very clearly with a division of δυνάμεις, and apparently it is supposed to be a complete division of all possible forces: αἱ μέν—αἱ δέ. And what is the criterion for the division? Put concisely: a major difference of the realms of beings to which each of the forces belongs; the expression ἐνυπάρχουσιν does not simply mean the coming forth and emerging for a time of this or that force in this or that region, but this concise Greek form of expression holds the meaning: the forces belong to and co-constitute the being and the being character of the realms of being that are encountered; accordingly, the division of forces can be accomplished by following the division of the realms of being. The result is a division of forces into δυνάμεις ἄλογοι and δυνάμεις μετὰ λόγου, forces without discourse and forces directed by discourse; this same division for the corresponding δυνατά is found at b22f.: τὰ ἄνευ λόγου δυνατά—without discourse, and τὰ κατὰ λόγον δυνατά—in keeping with discourse (likewise later on at 1048a3ff.).

This division ensues in accordance with the division of the realms of beings into those without souls and those that are besouled. And yet, the two divisions do not readily coincide. In order to explore this and to glean from it the significance of the division for the task of the essential determination of δύναμις κατὰ κίνησιν, it is necessary to discuss, even if only briefly, several points.

First of all, the divisions and the dividing. The splitting of beings into those without soul and those that are besouled appears to be clear; but it seems also that it is as empty of significance as any negative division by way of bifurcation, dichotomy. This kind of division in fact has the advantage of being in each case correct and complete; for example, the division of the whole body of human beings on our planet into those possessing motorcycles and those not possessing them is just as complete as it is correct. We observe immediately that this division does not tell us much because we can produce at will endless others of this sort. If we do not restrict the realm of division to human beings and extend it instead to all beings, then it appears that, with regard to this, the division into those without soul and those besouled, and non-living and living has the same dual character as the aforementioned division of all human beings: beings that do not possess soul, and those that do possess it—those without soul, ἄψυχον, and those that are besouled, ἔμψυχον. The manner of division, the form, the formula is the same; only the what of what is divided is different.

And yet, entirely apart from the character and the scope of the divided realms, this last division into two realms of being is different from the division into those human beings who possess motorcycles and those who do not possess them. Although, formally, in both cases something is presented as a negative bifurcation, it is still possible to ask how that upon which the division occurs is related to what is divided. Thus, the motorcycle is indeed a determination of a human being, but it is not essential for the human being; he can also exist without the motorcycle. Beings that are besouled can in no way be the beings that they are without soul. But is the same not true of a

§ 13. Concerning λόγος and soul

human who possesses a motorcycle? Surely this person also cannot be such a human without being a human who possesses a motorcycle. True, but nevertheless he can, if he does not possess a motorcycle, still continue to be that which he already is, whether he possesses a motorcycle or not, namely, a human being; thus that which he must be in the first place in order to be able to possess something like a motorcycle. Even though in both cases the division is made on the basis of the possessing and not possessing of something, nevertheless in the first case the nonpossession (of the motorcycle) does not involve the being of the possessor; what is there in possession does not represent any essential determination of what is divided, human beings. In contrast, the second division is made on the basis of the possession and nonpossession of something of the sort that absolutely belongs to the constitution of the being of the beings in question. Accordingly, the not possessing, the remaining withdrawn, is not unessential; the negation is an essential negation. But one could still say that if a being does not possess a soul, it can nevertheless remain and be a being. Certainly. However, what it is then is precisely no longer what it was as a being. Were the first case to hold, then the human through the loss of the motorcycle would need to become an animal or some other similar kind of being. Many indiscriminate things can be implicit in the human's lack of a motorcycle; for example, that the one in question perhaps does not have the needed money to procure it, or that he does not need a vehicle of this sort, or that he finds such a thing altogether tasteless, or that it never even crosses his mind to consider whether to possess or not possess it. On the contrary, by not possessing a soul, something very determinate is given. Through being besouled, a being is something living. "Life," however, is a way of being; if therefore this way of being is altogether lacking, but the one that lacks it is nonetheless addressed as being, then this points to those determinations which necessarily define this lifeless being as a being in its being in contradistinction to something alive. We notice that the lifeless is something other than the dead; a stone is never dead, not because it continuously lives, but because it does not live in the first place and therefore cannot at all know death. The second division is, although formally the same as the first example, nevertheless of a different character; it says more. This means it requires essentially more for its enactment, namely the glimpse into the being constitution of the being in question as such.

In all of this we have juxtaposed simply for our present purposes two examples of division. These deliberations appear trivial, and yet behind them lies hidden a decisive problem. What the division of beings into essential realms concerns, in contrast to the usually carried out division of a region of objects—and this is just what "realm" means—encompasses peculiar questions which we in philosophy have heretofore taken much too lightly; and thus we have adequately clarified neither the essence nor the proper ground of the division carried out here.

Hence if Aristotle refers the division of δυνάμεις back to such an essential division of the realms of being, thus back to a division of the ways of being, and if—as we heard—the δυνάμεις as such belong in each case to the essence of being, then we learn at the same time by this division of the δυνάμεις something essential about δύναμις in general and its essential possibilities. But then the discussion of the second chapter is to be brought to the same level as that of the first. We should attend no less, therefore, to what yields itself essentially with regard to δύναμις κατὰ κίνησιν in general, than to its modifications in the said realms of being.

*

A division may not be imposed externally onto what is to be divided; as something which belongs to the being of the divided, that upon which the division occurs must be derived from what is to be divided. So what about the division of δυνάμεις into those without discourse and those directed by discourse? Moreover, how does this twofold division relate to the twofold division of beings into those without souls and those with souls? This appears to be a matter of a simple congruence, yet this relation is by no means immediately clear, as we shall see.

We begin with the first question: What does δύναμις ἄλογος or μετὰ λόγου mean? What does λόγος mean here? We translate it as

§ 13. Concerning λόγος and soul

"discourse." But just what does discourse have to do with force, such that there comes to be with reference to λόγος an essential division of δυνάμεις?

We treated the original meaning of λόγος at the very beginning of this lecture. Λέγειν: to glean, to harvest, to gather, to add one to the other, and so to place the one in relation to the other, and thus to posit this relationship itself. Λόγος: the relation, the relationship. The relationship is what holds together that which stands within it. The unity of this together prevails over and rules the relation of what holds itself in that relation. Λόγος means therefore rule, law, yet not as something which is suspended somewhere above what is ruled, but rather as that which is itself the relationship: the inner jointure and order of the being which stand in relation. Λόγος is the *ruling structure*, the *gathering* of those beings related among themselves.

Such a gathering, which now gathers up, makes accessible, and holds ready the relations of the related, and with this the related itself and thus individual beings, and so at the same time lets them be mastered, this is the structure we call *"language,"* speaking; but not understood as vocalizing, rather in the sense of a speaking that says something, means something: to talk of or about something to someone or for someone. Λόγος is *discourse*, the gathering laying open, unifying *making something known* [*Kundmachen*]; and indeed above all in the broad sense which also includes pleading, making a request, praying, questioning, wishing, commanding, and the like. One mode of discourse understood in such a broad sense is the simple assertion [*Aus-sage*] about something, whereby discourse accomplishes this: It makes known in an emphatic sense that of which and about which the discourse is, and simply lets it be seen in itself. But questioning too is a making known in the sense of exploring [*Er-kunden*]; prayer is a making known in the sense of witnessing and attesting to, and likewise with the wish, or the refusal as when we decline, deny, and so on. Public discourse is also an announcing [*Ankündigung*], proclaiming [*Verkünden*], and declaring [*Künden*]. Λόγος is thus discourse in the utterly broad sense of the manifold making known and giving notice [*Kundgeben*]—*"conversance"* [*Kundschaft*].

The current translations of λόγος as "reason," "judgment," and "sense" do not capture the decisive meaning: gathering joining and making known. They overlook what is originally and properly ancient and thus at once essential to the word and concept. Whether, then, in the history of the origin of the word λόγος the meaning of the gathering joining was immediately accompanied by the meaning of gathering saying, a meaning that language always already has assumed, and in fact in the manner of conversance; whether, in fact, originally language and discourse was directly experienced as the primary and genuine basic way of gathering joining, or whether the meaning of gathering and joining together was only subsequently carried over onto language, I am not able to decide on the basis of my knowledge of the matter, assuming that the question is at all decidable. (In any case, we already find within philosophy the multiplicity of the meanings of λόγος in Heraclitus.)

What then does the determination of ἄλογος and μετὰ λόγου mean now on the basis of this clarification of the word λόγος? Ἄλογος: without discourse: without conversance; this means something which is without conversance in what and how it is. Without conversance: without the possibility of taking notice, of perceiving, or of giving notice, and hence all the more not being in a position to explore and be conversant in a matter. Μετὰ λόγου, in contrast, is something which has conversance there along with it in what and how it is. Conversance: the possibility of taking and giving notice and thus the possibility of exploring and becoming conversant and so being conversant.

Thus Aristotle divides, as we heard, the δυνάμεις κατὰ κίνησιν into what is without discourse and what is directed by discourse, *without conversance* and *conversant*. And in fact he achieves the division by going back to a division of beings into ἄψυχα (soulless) and ἔμψυχα (besouled). Thus ἄλογον corresponds to the ἄψυχον and μετὰ λόγου to ἔμψυχον. In general, λόγος and ψυχή, soul, correspond. Where λόγος, there ψυχή, and where ψυχή and ἔμψυχον, there λόγος and μετὰ λόγου. Or is this last relationship simply invalid? Let us see.

In what regard are ἄψυχον and ἔμψυχον different? Aristotle says in *De anima*, B 2, 413a20ff.: λέγομεν . . . διωρίσθαι τὸ ἔμψυχον τοῦ

§ 13. Concerning λόγος and soul

ἀψύχου τῷ ζῆν. "We say that the besouled is differentiated and delimited over and against what is without soul by life, being alive." But this account does not say much because Aristotle must remark immediately: πλεοναχῶς δὲ τοῦ ζῆν λεγομένου. "Life is understood in many ways." What makes something a living being, what determines it essentially as being in this way, can be many things. Plants, animals, humans all mean beings that live, and nevertheless their way of being is different. So, for example, neither plants nor animals, although defined as ζωή, have any βίος, life in the sense of a life history; that is, they do not have the possibility of a freely chosen and formed Dasein that holds itself in what we call composure—and, accordingly, neither do they have the possibility of an uncomposed, haphazard life. In spite of this, it can and must be asked whether a differentiation is not indeed established between those things which are in the mode of life—irrespective of the differentiation within what lives—and those things which are in the mode of the lifeless.

In fact, Aristotle offers such a distinction, and even in the same treatise, and tells us at the same time what distinguishes the way of being in the sense of *life* as such (*De an.* A 2, 403b25f.): τὸ ἔμψυχον δὴ τοῦ ἀψύχου δυοῖν μάλιστα διαφέρειν δοκεῖ, κινήσει τε καὶ τῷ αἰσθάνεσθαι. "The besouled appears to be distinguished from the soulless first and foremost by two characteristics, movement and perceiving." Κίνησις here means the *self-moving*, not only in the sense of change of place but in the sense of taking nourishment, of growth and degeneration. But what distinguishes the ζῆν μόνον, the just-living-there or "vegetating," over and against the ζῷον, the living being in the sense of animals, is just this: that the "just-living" lacks what is proper to the ζῷον, namely ἡ αἴσθησις (πρώτως)—*perceiving*.

What lives is always surrounded, related to its surroundings, where what surrounds presents itself in some way or another, and in such a way that, admittedly, its determination presents the utmost difficulty and is exposed to the danger of being overdetermined. The basic form of αἴσθησις, the relationship to the surroundings as such, is ἁφή, touching, feeling, grasping (and likewise the other forms up to ὄψις).

Plants and animals are therefore besouled, ἔμψυχα. The ζῷον even has the possibility of taking what presents itself in the surroundings;

it has τὸ κριτικόν (Γ 9, beginning): the possibility of the *separating out* and *bringing out* of something, for example, to stalk prey, to lie in wait, to notice, to know their dwelling places, to protect itself against attackers, and so on. Therefore, exploring is proper to the animal. Is the animal then μετὰ λόγου? Surely not, for it is the distinguishing definition of the human over and against the animal that it is ζῷον λόγον ἔχον—that animal which has conversance at its disposal. According to this, then, the animal is ζῷον ἄλογον, but as ζῷον nevertheless ἔμψυχον, therefore not ἄψυχον. And so the division between ἄψυχον/ἔμψυχον does not coincide with ἄλογον/λόγον ἔχον. Therefore there is also within the ἔμψυχα the besouled forms of life, ἄλογα (plants, animals).

If ψυχή is there, this does not mean that λόγος is there. This is in accord with what Aristotle says in the *Nicomachean Ethics*, Z 2, 1139a4: δύ' εἶναι μέρη τῆς ψυχῆς, τό τε λόγον ἔχον καὶ τὸ ἄλογον. "There are two parts to the soul, one defined as discourse directed, and one that is without discourse." This is a differentiation within the besouled known already by Plato. If, however, the realm of the ἄλογον extends itself into the realm of the ἔμψυχον, then a double meaning enters into the concept of ἄλογον: both the stone and the rose are ἄλογον. See *Met.* Θ 5, 1048a3ff.; here it is explicitly stated: τὰ ἄλογα . . . ἐν ἀμφοῖν—beings without discourse, without conversance, are found in both the soulless and the besouled. But stone and rose are ἄλογα in different senses, and here we can directly apply what we learned about the various ways of στέρησις.

We now see that λόγον ἔχον is in fact necessarily an ἔμψυχον, but not every ἔμψυχον is necessarily a λόγον ἔχον. But here the difficulty arises again that we already touched upon. Αἴσθησις (the κριτικόν) belongs to the essence of being an animal (animality). Is this not already a kind of λόγος, conversance? So in the end is the animal not indeed ζῷον λόγον ἔχον? But over and against this stands the fact that this determination is precisely the essential definition of the human. This shows that the question of whether the animal does not also have λόγος, on the basis of having αἴσθησις, can emerge only if we comprehend λόγος as conversance, instead of relying on the well-known and reductive conception and translation of λόγος as

§ 13. Concerning λόγος and soul 107

reason. If we do this, then everything becomes clear in one fell swoop. The animal may indeed have a certain kind of exploring and perceiving, but nevertheless it remains without reason, in contrast to humans, who are animals with reason. As Kant formulated it: The cow cannot say "I," it has no self. To be sure, in this way everything becomes clear and simple—but the question remains whether we thereby stay close to the core of the Aristotelian posing of the question, and whether we thereby adhere to the original ancient content of the concept of λόγος. With this we disregard entirely the difficulty of having to say what reason means here, and in what sense "reason" is to be understood. We must above all adhere to what Aristotle presents as fact: that indeed the animal is αἰσθητικόν, κριτικόν—in the manner of bringing out. And just as little are we allowed to shove aside the developed meaning of λόγος in the sense of conversance. For the matter surely demands that we do not deny λόγος to the animal as it now stands—or else leave the question open. And this is just the position that Aristotle takes unambiguously at *De an.* Γ 9, 432a30f.: τὸ αἰσθητικόν, ὃ οὔτε ὡς ἄλογον οὔτε ὡς λόγον ἔχον θείη ἄν τις 'ῥᾳδίως. "No one may easily settle, with regard to the ability to perceive, whether this is a capability without conversance or a conversant capability." This caution with regard to deciding and questioning must even today remain for us exemplary, irrespective of the further question of where the essential boundary runs between animal and human.

Λόγος does not mean reason. The Aristotelian problem makes sense only if λόγος has a certain kinship to αἴσθησις. This kinship lies in the fact that both—the exploring and being-conversant as well as the perceiving—in some way uncover and unconceal that toward which they are directed. Both αἴσθησις and λόγος are connected with ἀληθεύειν (which at first has absolutely nothing to do with knowledge in the sense of theoretical comprehension and intention).

The extent to which Aristotle also intends in a certain sense to ascribe to animals λόγος—conversance in the sense of a circumspection which knows its way around—can be seen in *Met.* A 1, where Aristotle attributes to some animals the possibility of φρονιμώτερον and thus a certain φρόνησις (something like circumspection)

(980b21). Here it should be noted that besides ethical and practical behavior, φρόνησις also signifies the self-sensing of human beings. On this point, I am leaving aside the difficult passage (*De an.* B 12, 424a26ff.) where αἴσθησις is directly designated as λόγος τις. We should understand λόγος in this passage neither merely as relationship, nor simply as reason or discourse in the sense of language; rather, what is in fact meant by the λόγος τις is the perceiving exploration of . . . , and the conversant relating to . . . , the relation which takes cognizance of its surroundings, the relation to what presents itself in the surroundings as lying opposite, as ἀντικείμενον.

We have thereby clarified, to the extent necessary for us, the relationship of the two divisions: ἄψυχον—ἔμψυχον and ἄλογον—λόγον ἔχον. Now let us return to our text (*Met.* Θ 2, 1046a36ff.). We are now in a position to read with more precision and to observe that Aristotle has already taken into account everything just said. With αἱ μέν—αἱ δέ, he is not simply setting apart ἄψυχα and ἔμψυχα, soulless and besouled; rather, he defines more closely in what respect he means ἔμψυχα when he says: καὶ ἐν ψυχῇ (a37), and indeed the besouled, that is, the besouled body taken only according to its besouledness ("in the soul as such"). Thus, the bodily is thereby excluded. This is in fact not identical with corporeality in the sense of the constitution of a material thing of nature, but it nevertheless displays processes, for example physico-chemical processes, which are able to be grasped within certain limits without observing the besouledness. In this excluded realm, which nonetheless belongs to the besouled, there is ἄλογα. But not even ἐν τῇ ψυχῇ is an unequivocal determination; besouledness is also the specific life form of plants, which (although they have soul) are always nevertheless ἄλογον. Only when the besouled in its besouledness is taken in an entirely different way—as the besouled being that has λόγος—only then is the ἔμψυχον the opposite of the ἄλογον.

When we speak of the besouled being who has λόγος, we do not mean that λόγος, conversance (discourse), is merely added on; rather, this ἔχειν, having, has the meaning of being. It means that humans conduct themselves, carry themselves, and comport themselves in the way they do on the basis of this having. The ἔχειν means having in

§ 13. Concerning λόγος and soul

the sense of governing over . . . ; to be empowered for conversance and above all through conversance (λόγος) means: to be conversant in oneself and from out of oneself.

This λόγον ἔχον is again doubled (*Nic. Eth.* Z 2, 1139a12) into the ἐπιστημονικόν and the λογιστικόν; ἐπιστήμη means a versatile understanding of something, being familiar with something and having knowledge of it; λογισμός means circumspective calculation and deliberation. It is therefore related to choice and decision. Both belong to λόγος as conversance, on the basis of which human beings are aware of things and investigate them. At the same time, they are aware of their own possibilities and necessities. Whenever this conversance addresses itself to things and discusses them, it is a conversance which deliberates with both itself and others; a conversance which debates with itself and calls itself into account. It is an "I" saying. "Language" is understood here in the broadest sense of λόγος as a conversant gathering, as a gatheredness of beings in "one"; in Dasein, which is at the same time a dissemination.

This is our understanding of the definition: the human being is ζῷον λόγον ἔχον—the living being who lives in such a way that his life, as a way to be, is defined in an originary way by the command of language. The original understanding of language, which was of fundamental importance for the definition of the essence of human being, gains expression in Greek in such a way that there is no word for language in our sense. Rather, what we call "language" is immediately designated as "λόγος," as conversance. The human being "has the word"; it is the way he makes known to himself his being, and the way in which he sees himself placed in the midst of beings as a whole (compare Plato, *Cratylus* 399c). To be empowered with language—; language, however, not merely as a means of asserting and communicating, which indeed it also is, but language as that wherein the openness and conversance of world first of all bursts forth and is. Language, therefore, originally and authentically occurs in poetry [*Dichtung*]—however, not poetry in the sense of the work of writers, but poetry as the proclamation of world in the invocation of the god. But nowadays we see language primarily from the point of view of what we call conversation and chitchat; conventional philology is in accord with this.

*

Chapter 2 begins with a differentiation of δυνάμεις. This difference is a pervasive theme pursued with the aim of procuring thereby an originary understanding of δύναμις κατὰ κίνησιν in general. The division ἄλογον and μετὰ λόγου is brought back to the division of ἄψυχον and ἔμψυχον. We obtained the following clarifications: (1) Λόγος—from λέγειν: to gather, to bring into relation and relationship—means relationship, relation, rule, law, ruling framework, "language" in the sense of discourse, conversance. (2) The difference between ἄλογον and μετὰ λόγου accordingly means: without conversance and conversant. (3) The difference that this division was based upon, namely ἔμψυχον—ἄψυχον, was reached in regard to ζῆν (living being), and even more plainly by the characteristics of κίνησις, and of αἴσθησις and κριτικόν (through them the ζῆν μόνον is also able to be distinguished from the ζῷον). Αἴσθησις as λόγος τις is a relationship to the surroundings that makes known and informs. It is not easily determined whether animals are ἄλογον or λόγον ἔχον. Saying that αἴσθησις is a relationship to the surroundings, a taking cognizance, does not say that what makes itself known there is perceived as being [*Seiendes*]. But the human being is ζῷον λόγον ἔχον, the living being for whom language is essential; or better, the being for whom discourse is essential, discourse understood in its original sense of expressing oneself about the world and to the world in poetry. We can infer from this concept of λόγος what "logic" means, namely a philosophical knowledge of λόγος—something very different from what we usually understand by logic, whether it be formal or transcendental logic.

We ourselves as human beings are the beings who are μετὰ λόγου in the authentic sense, that is, the beings who exist. Therefore, in the text (*Met*. Θ 2, 1046a37), the double καί, καί after ἐν τοῖς ἐμψύχοις signifies a progressive restriction, and, along with this, an explanation of what is meant by ἔμψυχον. Aristotle thereby concedes that the domain of ἄλογον overlaps that of ἔμψυχον, and that this domain is not identical to that of μετὰ λόγου.

§ 14. The extraordinary relationship of force and conversance in δύναμις μετὰ λόγου, in capability

The division into ἄλογα and ὄντα μετὰ λόγου should then include within itself such a division of the corresponding δυνάμεις. To speak more precisely: That which is μετὰ λόγου, λόγον ἔχον (presiding over . . .), is already in itself empowered toward something, so much so that even this being empowered toward . . . is what it is only in that conversance belongs to it. *Conversant force*—δύναμις μετὰ λόγου; we choose the word *capability* [*Vermögen*] to express this mode of forces.

Aristotle says at 1046b2-4: διὸ πᾶσαι αἱ τέχναι καὶ αἱ ποιητικαὶ ἐπιστῆμαι δυνάμεις εἰσίν· ἀρχαὶ γὰρ μεταβλητικαί εἰσιν ἐν ἄλλῳ ἢ ᾗ ἄλλο.

"For this reason, all skills and ways of versatile understanding in the production of something are forces (thus capability in our sense); for they are that from out of which, as in another, this is directed toward an ability to shift, a transformability."

Ποιητικὴ ἐπιστήμη is a versatile understanding of ποίησις, an understanding of producing and work, and not just ἐπιστήμη alone, not a mere familiarity and acquaintance with things. Such a familiarity with things does not try to make them, but lets them be what they are, solely for the sake of investigating them and being knowledgeable about what they are and how they are. This kind of conversance is science; ἐπιστήμη ποιητική, on the other hand, is τέχνη (see *Nic. Eth.* Z 3-4). But τέχνη can also have the meaning of a pure being familiar with things. We can gather from all this that the Greek concept of knowledge in general is essentially determined in this way, that is, in terms of the human being's basic relation to the work, to that which is fulfilled and fully at an end. Of course, this has nothing to do with a primitive understanding of the world which operates within a horizon of handmade artworks instead of our supposedly higher mathematico-physical horizon. We will gain greater clarity in our understanding of the inner relationship of all the Greek concepts of knowledge and of the essential relation of λόγος to the work by inquiring further about the relationship of δύναμις and λόγος.

We name, then, these δυνάμεις, which in fact are ἔμψυχα but nevertheless ἄλογα, and so an intermediate level between ἄλογα and λόγον ἔχοντα, capacities [*Fähigkeiten*]. Although these distinctions and determinations of corresponding names may be very useful, they remain empty and dangerous as long as they are not carried out with the corresponding understanding of the matter. And so the question is therefore once again raised whether here Aristotle merely wants to divide realms and enumerate kinds of forces or whether he has other intentions. In fact, the latter is true.

Even a cursory reading reveals that in what follows λόγος is constantly under discussion and, in fact, in relationship to δύναμις. The guiding aim is to make more poignantly visible the essence of δύναμις by elucidating the extraordinary relationship between δύναμις and λόγος, and, above all, to prepare a question which has an inner connection with the question concerning δύναμις κατὰ κίνησιν, namely the question concerning ἐνέργεια κατὰ κίνησιν (compare *Met*. Θ, chaps. 3-5). This relationship of λόγος and δύναμις obtains in this way its clarification, that δύναμις μετὰ λόγου is constantly contrasted with δύναμις ἄλογος. Δύναμις μετὰ λόγου itself requires a new discussion of the already touched upon connection between δύναμις and στέρησις (compare p. 131ff.).

a) Capability necessarily has a realm and contraries that are in that realm

We will divide the following considerations into single steps:

1046b4-7: καὶ αἱ μὲν μετὰ λόγου πᾶσαι τῶν ἐναντίων αἱ αὐταί, αἱ δ'ἄλογοι μία ἑνός, οἷον τὸ θερμὸν τοῦ θερμαίνειν μόνον, ἡ δ'ἰατρικὴ νόσου καὶ ὑγείας.

"And indeed forces which are in themselves conversant are always, as the same, directed at contraries. However, those without conversance are, as one, directed at a singular. For example, the warm is directed only at making warm, but the art of doctoring is directed at sickness and health."

The text now makes clear what is meant by this character of conver-

§ 14. *Relationship of force and conversance* 113

sance which δύναμις has, and what consequences this has for the essence of δύναμις. This kind of force which is inherently conversant (μετὰ λόγου) carries with itself the implication that it is directed at more than that which is without such awareness. The realm in which it reigns is wider. The one is directed only at warmth and the realm of warmth, the other at sickness and also health. Indeed, one might wonder whether this separation of making warm from healing holds up. For the art of doctoring is for the most part wrapped up in matters which concern being sick. Were it not for sickness, we would not need doctors; and health appears through the removal of sickness. But the same relationship exists with making warm, which goes out of a warm body and into another. Through the transference of warmth from one body to another, cold is displaced. Warmth is as much the displacing of cold as is healing, which is the bringing forth of health, the displacement of sickness. Either we must say that warming also unfolds from a twofold (like healing), or we must say that the art of healing really unfolds only from a singular (like warming), from sickness. There is therefore no difference in the reach of the realm of both powers. And therefore Aristotle is mistaken. But does Aristotle at all want to say what we were attempting to refute by the above considerations? Is it simply a matter of becoming aware that with δύναμις μετὰ λόγου in relation to what it is directed at, only more occurs than happens with δύναμις ἄλογου? Or should it not rather be shown that in δύναμις μετὰ λόγου the contrary of that to which it is related also occurs? But why is the contrary, the displacing and disappearing of cold and thus cold itself, concealed in the case of warming? The contrary of warming does indeed play a role here. So with what right does Aristotle say: ἡ δύναμις μετὰ λόγου τῶν ἐναντίων, ἡ δύναμις ἄλογος ἑνὸς μόνον? "The force which is in itself conversant, capability, is directed at contraries, whereas that which is without conversance is directed only at one." But let us take a good look at this. Aristotle says nothing of the sort. Rather, his thought is: ἡ δύναμις μετὰ λόγου τῶν ἐναντίων ἡ αὐτή and ἡ δύναμις ἄλογος μία ἑνός. "The force which is in itself conversant is directed, as the one and the same that it is, at contraries (that is, the one and its opposite other); that which lacks conversance, however, is directed, as the one that it is, only at one."

Therefore the text is not about the greater or the lesser extent of the realm, not about whether the contraries play a role or not; rather, it is about the fact that this orientation of the art of doctoring as a healing of sickness is already in itself and in fact necessarily oriented toward health. On the other hand, the warmth which goes out of a hot body—this giving away of warmth to another—need not necessarily nor in advance be oriented toward cold and its disappearing. Of course, this distinguishing of both of the δυνάμεις does concern a differentiation of their realms; but it is not so much a matter of the largeness or smallness of the realms; rather, it mainly involves the way in which both realms of force are given and how this givenness of the realm belongs to the essence of force. What gives δύναμις μετὰ λόγου a special significance is that its realm is given to it necessarily and completely according to its ownmost potentiality; whereas for δύναμις ἄλογος, the realm not only remains closed off, it lies completely outside the possibility of being opened up or closed off. Yet we cannot even simply say that the realm is completely lacking. The orientation of a hot body, as warming, is not arbitrary. It is, for example, not oriented toward relations between numbers, much less toward a proposition of science or the like.

And so, because the openness of the realm of force happens for this force in and through the λόγος which belongs to it, the open realm is not only completely wider, but within this realm the contrary is necessarily posited, in the relational realm of force. This is made clear in the next passage, 1046b7-15:

αἴτιον δὲ ὅτι λόγος ἐστὶν ἡ ἐπιστήμη, ὁ δὲ λόγος ὁ αὐτὸς δηλοῖ τὸ πρᾶγμα καὶ τὴν στέρησιν, πλὴν οὐχ ὡσαύτως, καὶ ἔστιν ὡς ἀμφοῖν, ἔστι δ' ὡς τοῦ ὑπάρχοντος μᾶλλον. ὥστ' ἀνάγκη καὶ τὰς τοιαύτας ἐπιστήμας εἶναι μὲν τῶν ἐναντίων, εἶναι δὲ τοῦ μὲν καθ' αὐτὰς τοῦ δὲ μὴ καθ' αὐτάς· καὶ γὰρ ὁ λόγος τοῦ μὲν καθ' αὐτό, τοῦ δὲ τρόπον τινὰ κατὰ συμβεβηκός. ἀποφάσει γὰρ καὶ ἀποφορᾷ δηλοῖ τὸ ἐναντίον· ἡ γὰρ στέρησις ἡ πρώτη τὸ ἐναντίον, αὕτη δ' ἀποφορὰ θατέρου.

"The reason for this (that certain δυνάμεις are directed at contraries) is that understanding something is (in itself) a conversance (cognizance). However, this inquiring conversance, that is, the one and the same, discloses the everyday things with which we deal and their

§ 14. Relationship of force and conversance

withdrawal; admittedly not in the same way; that is, in a certain respect the exploration does have to do with both; but in another respect it has more to do with what is (always already) there in advance. Hence the necessity that the so-constituted (λόγος-directed) ways of versatile understanding of something refer on the one hand to contraries (the one and its other), and on the other hand to one of the contraries from out of itself (immediately, according to its orientation), and to the other not in the way that has been indicated. For cognizance also is directed at the one in itself, and at the other to some extent only incidentally; that is, through denial and removal it makes manifest the contrary; for the contrary[1] is that which is withdrawn in the primary sense, but this is the carrying away of the other (opposite to the one)."

What Aristotle says here offers at first no particular difficulties. Aristotle traces back to λόγος the manifestness and with it the given contrary, to which certain forces are related. Conversance is not only the abode of manifestness; it is also at the same time the site of the manifestness of the contraries. This is then once again explicitly applied to ἐπιστήμη μετὰ λόγου in such a way that the already touched upon character of λόγος and the essence of the contraries connected with it once again enter into the discussion. All this is done in a few clear steps in the presentation. And yet behind these steps there is hidden a far-reaching complex of essential philosophical questions to which the thinking of antiquity slowly brought the first light and rendered tangible distinctions.

When we examine the content of our passage, we discover first of all the connection of δύναμις κατὰ κίνησιν and λόγος; then the connection of both of these with στέρησις; and finally, the connection of these with the negative and the opposite, with opposition and the not.

We need far-reaching deliberations to see through to some extent the inner connection and common root of the questions here touched upon. It requires also a thoroughgoing interpretation of other Aristotelian treatises in order to see at the same time how the problematic

---

1. See Bonitz's *Commentarius*, p. 383; τὸ ἐναντίον is the explanatory subject.

of antiquity sustained all these questions on a very determinate level, and thereby gave a very determinate destiny to what was later and is today (and also in Hegel) meant by "logic." We shall here say only what is immediately required for the elucidation of the treatise on δύναμις.

b) The capability of producing: λόγος as innermost framework

First of all, about λόγος, which clearly stands at the center of Aristotle's considerations. We have already tried several times to get closer to the essence of the phenomenon that is designated in this way. It continues to be the most productive if what we call conversance is established as the most essential character of λόγος. We have here, however, undoubtedly hit upon a narrower meaning of the word λόγος, and it is a question of how what is meant by this is connected to the fundamental phenomenon. In what sense is the meaning of λόγος that we have come upon a narrower meaning? Λόγος is brought into relation to ἐπιστήμη and vice versa. It is stated: versatile understanding of something is λόγος; consequently, λόγος is also related to other things, not only to ἐπιστήμη. The question arises: in what form does λόγος become manifest in ἐπιστήμη, and in fact ἐπιστήμη meant as ποιητική (producing something)?

The thesis runs as follows: ἐπιστήμη ποιητική, as always one and the same, not only is directed at a singular, but, precisely as one and the same, in accordance with its essence (necessarily), it is directed at the one and the other: τῶν ἐναντίων. Why? Because ἐπιστήμη ποιητική is λόγος. To what extent is ἐπιστήμη ποιητική—λόγος? What, after all, is ἐπιστήμη ποιητική? To what extent can one say of it that it is directed at contraries? What does this mean? It means this: Producing indeed is always directed at one, what is to be produced (the shoemaker makes shoes and not pots), but in a way that, along with it, the contrary is taken into account.

We can first get closer to the facts in general by referring to an example such as pottery. The entire process of producing mugs, from the preparation of the clay, through determining the moisture of the clay and regulating the turning of the wheel, up to watching over the

§ 14. Relationship of force and conversance

kiln, is, so to speak, interspersed with alternatives: this, not that; in that way and not in another way. Production, in the way of proceeding that is appropriate to it, is in itself a doing and leaving undone—a doing something and leaving its contrary alone. Particularly because producing is in itself a doing and leaving alone, therefore, that to which it is related is ἐναντία.

But now we have to focus more sharply on this roughly formulated connection between production (ἐπιστήμη ποιητική) and what is related to it as contrary. That is, we have to grasp it out of the inner constitution of the essence of production. Not only did the Greeks, Plato and Aristotle, carry out the interpretation of this phenomenon of production, but the basic concepts of philosophy have grown out of and within this interpretation. (We will not discuss here why this is so and what it all means, or why ancient philosophy nevertheless was not just the philosophy of shoemakers and potters.)

*

What the Greeks conceived as ἐπιστήμη ποιητική is of fundamental significance for their own understanding of the world. We have to clarify for ourselves what it signifies that man has a relation to the works that he produces. It is for this reason that a certain book called *Sein und Zeit* discusses dealings with equipment; and not in order to correct Marx, nor to organize a new national economy, nor out of a primitive understanding of the world.

What then is ἐπιστήμη ποιητική, production? What is produced, what is intended for production, is the ἔργον. This does not result arbitrarily and by chance from any work or activity whatsoever; for it is always that which is intended to stand there and be available, that which must appear in such and such a way and offer this specific look. Indeed, how the work is to appear, its outward appearance, must be seen in the production and for it. The outward appearance, εἶδος, is already seen in advance, and this is so not only in a general and overall fashion; rather, it is seen precisely in what it comes to in the end, if it is to be fully ended and finished. In the εἶδος of the ἔργον, its being-at-an-end—the ends which it encloses—is in advance already anticipated. The εἶδος of the ἔργον is τέλος. The end which finishes, however, is in its essence, boundary, πέρας. To produce something is in itself to forge something into its boundaries, so much so that this being-enclosed is already in view in advance along with all that it *includes and excludes*. Every work is in its essence "exclusive" (a fact for which we barbarians for a long time now lack the facility).

We now have to see more clearly where this exclusiveness has its origin and how it extends into the whole complex of events and thereby into the essential constitution of production. For only when we have examined the extent to which producing a work is in itself excluding will it become clear just why and in what way producing is essentially related to a contrary, to what is excluded.

Producing is limiting and excluding primarily because the whole event of producing is, so to speak, secured to the anticipated outward appearance of the ἔργον as εἶδος, τέλος, πέρας. But how then does the exclusiveness which is situated here make itself felt? First of all, and in its predominant meaning, in that the εἶδος is in itself assigned to very definite material (ὕλη) as that out of which something is to be produced. A saw for sawing wood cannot be made out of anything whatever, for example, but must be made out of something such as metal. Insofar as producing is always producing something *out of* something, and insofar as this "out of which" is ever defined only by and in the exclusion of other things, boundaries extend forth in the producing itself.

However, producing not only involves material which does not come into play; it likewise involves precisely that material which is suitable. For material as such, for example, as iron, as metal, is precisely not yet what is to be made out of it. Seen from εἶδος and τέλος, it is, on the contrary, ἄπειρον, that which is without boundaries, the unbounded, that which has not yet been brought into bounds but, at the same time, is to be bounded. Precisely because the definitely demarcated material is tailored on the basis of the ἔργον, precisely for this reason, it likewise stands as unbounded over and against the εἶδος. Both are directed away from one another and yet toward one another; thus there is an opposition, and that is to say, a facing one another which is necessarily mutual—a neighborhood, and indeed one

§ *14. Relationship of force and conversance* 119

whose extension is the farthest. This is the concept of the Greek ἐναντίον: a lying opposite each other and confronting each other face to face; ἐναντιότης (contrariness), which Aristotle actually first fully clarified in its essence, is not simply what lies apart and is merely different yet of no concern to anything; rather, it is what lies over and against. Εἶδος, as τέλος and πέρας, necessarily furnishes itself with such an opposite as ἄπειρον; in the bounded ἄπειρον (of ὕλη), εἶδος becomes its μορφή. *Forma—materia*, nowadays this is a worn-out schema in philosophy, but it did not fall from the skies to be manipulated at will. Because this neighborhood of εἶδος and ὕλη lies in the essence of producing, producing necessarily, at each step along the way, is constantly excluding and enjoining, fitting in and, at the same time, leaving out.

Thus, it has become clear in what ways ἐπιστήμη ποιητική is related to ἐναντία. Certainly; but this also happened without the least reference to λόγος. So we can now say: The λόγος which belongs to ἐπιστήμη ποιητική is related to contraries because ἐπιστήμη ποιητική in its essence is directed at ἐναντία. But then we have arrived at the exact opposite result from Aristotle; for Aristotle says the reverse: ἐπιστήμη ποιητική is related to ἐναντία because it is λόγος. This is the basis of the inner contrariness of producing, and conversely, the contrariness of λόγος is not a consequence of the essence of the ἐπιστήμη to which it belongs. Which position represents the truth? Or can both theses be reconciled? Should that be possible, what would be the situation with regard to the relation of ἐπιστήμη ποιητική and λόγος?

It cannot be doubted that our interpretation of the essential constitution of producing is correct. Moreover, it certainly corresponds to what the Greeks themselves believe about the relation of ποίησις, εἶδος, τέλος, and ὕλη. But just as little can it be disputed that Aristotle says unequivocally: ἐπιστήμη ποιητική is directed at contraries because it is λόγος. The thesis that λόγος as such is the ground and origin of ἐναντιότης is contained in this statement. We have seen, however, on the basis of the explanation of the essence of ποίησις that contraries reside in εἶδος. If both theses are to be reconciled, then, according to this, there must be an inner relation between εἶδος and λόγος, and this, in turn, in such a way that εἶδος is λόγος, and therefore the seat of contrariness and neighborhood. For only on this account is Aristotle's thesis that ἐπιστήμη, as λόγος, is directed at contraries preserved.

So what about the connection between εἶδος and λόγος? We have to try to decide this question without arbitrary speculation about concepts and words, but upon the very soil in which all these questions of ours arose: by constantly keeping in view the essential constitution of producing a work—not only as a fundamental comportment of man but as a decisive determination of the existential being of the Dasein of antiquity.

In order to comprehend, however, the inner connection between εἶδος and λόγος, we must first completely disengage ourselves from all the new interpretations and superficial meanings which have in the meantime been attributed to both of these words. Seen from the point of view of these meanings, the question we are troubling ourselves over is, of course, not a serious one. Everyone knows that λόγος for Aristotle means "concept" and εἶδος means "species." A species is a definite class of concepts which is distinguished from concepts of genus. Concept and species concept are essentially the same. So, what is there about the relation of εἶδος and λόγος that remains to question? This not only is convincing but corresponds to the philologically exact procedure of keeping to the facts. Translating εἶδος as "outward appearance" and λόγος as "conversance" is an example of the kind of unscientific procedure based on a certain philosophy which is currently fashionable. It reads back into antiquity contemporary viewpoints. Because historians of philosophy tell each other that λόγος means "concept" and because everyone believes it, and most of all because no one has anything in mind by this, naturally this translation is consistent with the facts. But what historical facts are is a matter all its own; and even more so is what we call a "historian." For a long time now we have believed that every clever writer—and who today does not write—that every writer who vents his opinions about the past is a historian. So we willingly admit that what we are doing here is historically false, that is, false according to the judgment of professional historians of philosophy. We now want only to understand one

§ 14. Relationship of force and conversance 121

thing: to what extent what Aristotle calls λόγος is connected with εἶδος, and to what extent λόγος is the basis of the fact that ἐπιστήμη is related to contraries.

In producing something, the thing to be produced must necessarily be previewed even though it is not yet finished or perhaps not even begun. It is simply represented (*vor-gestellt*), in the genuine sense of the word, but not yet brought about and produced as something at hand. This representing and previewing of the ἔργον in its εἶδος is the real beginning of producing and not, for example, mere making in the narrow sense of working with one's hands. This taking the outward appearance into view is in itself the forming of an aspect, the forming of a model. But we know something with regard to this: the formation of a model can occur only as a bringing into bounds of what belongs to the model. It is a selecting, a selective gathering of what belongs together, a λέγειν. Εἶδος is a kind of being gathered together and selected, a λεγόμενον; it is λόγος. Εἶδος is also τέλος—the ending end, τέλειον—the perfected, the fulfilled, the gleaned, the selected; τέλος is, in accord with its essence, always selected: λόγος.

But εἶδος is λόγος even in the meaning of λόγος which we at the same time misunderstand, when λόγος signifies discourse, language, saying. Εἶδος is what it is only insofar as along with it and through it something which is to be produced is addressed as what is to be present later. Selection is addressing as . . . , λέγειν. The "addressing as" or, more exactly, this "as" itself has the character of "as this or that." The as is always in some way or other a selecting with a view toward something.

The εἶδος says what is to be produced. It is the such and such which is addressed as this or that. It intrinsically excludes others. The εἶδος assumes leadership in the whole process of production. It is the authority and regulator which says what the standard is. It does so from out of itself—καθ' αὐτό (1046b13), but always in a way that excludes others. This other is, however, what is constantly present along with it. It is what occurs with it [*das Bei-läufige*]—κατὰ συμβεβηκός (b13) inasmuch as the material and each particular state in the course of production offer occasions for mistakes and failure and for being irregular. Thus λόγος, the selected and above all the addressed, is constantly what excludes, but this means that it is what includes the contrary with it. What this says is that the contrary is "there" and manifest in a peculiar way in the very fact of avoiding it and getting out of its way. Of course, it is not manifest by itself—that which is to be avoided is not what occupies the potter. Rather, it is manifest incidentally, not in the sense of by chance but in the sense of necessarily following along with something else. Accordingly, λόγος has to do with both πρᾶγμα and στέρησις, but of course not in the same way—οὐχ ὡσαύτως (b8-9).

In this way we can understand the extent to which λόγος is the origin of the ἐναντίον, what lies over and against. More precisely, we can see clearly the extent to which λόγος is the ground of the fact that the ἐναντία announce themselves as such in every production.

But an objection can be raised against this interpretation. Someone might perhaps wish to point out that Aristotle sees the matter far more simply and clearly. The ἐναντιότης is given with λόγος because λόγος is not only κατάφασις but also ἀπόφασις, not only affirmation but also negation. In other words, λόγος is judgment, and of course there are positive and negative judgments. Since ἐπιστήμη ποιητική is ἐπιστήμη, that is, knowledge, and since all knowing is by common consent judging, judgment, along with its two contrary forms, belongs to ἐπιστήμη and gives to ἐπιστήμη its relation to contraries. Logically this is absolutely correct, and this line of thinking is to many people not only convincing but even perspicacious. It has only the one disadvantage that it does not say anything and does not have anything in mind. This explanation explains nothing. It presupposes what is to be explained. For what do we mean by the statement which we find in every logic book: "There are positive and negative judgments"? Are these "given" in the same sense that there are both birds and vermin during the summer? And even supposing there are negative and positive judgments, why are negative judgments also elicited when producing something? They could surely be omitted, and then there would be only positive judgments, and so only one side of the ἐναντιότης. But this would mean there would be no contraries at all.

Why, then, does this division of affirmation and negation pertain to λόγος? This is the question we cannot evade if we want to get any

§ 14. Relationship of force and conversance 123

idea of the whole interrelation between δύναμις μετὰ λόγου and ἐπιστήμη ποιητική. Aristotle has this interrelation in mind as he further develops the δύναμις problem.

*

The inner relation of δύναμις and λόγος drew our attention to δύναμις μετὰ λόγου. What is characteristic of this is that it is directed at contraries. What does this mean, and to what extent does it characterize δύναμις μετὰ λόγου, that is, ἐπιστήμη ποιητική? Ἐπιστήμη ποιητική is a being familiar with the producing of something, with something in its producibility, or even better in its being produced, as ἔργον. Εἶδος, τέλος, and πέρας are determinative for this work relation—the forming of a model as a forging into bounds. We find here a preliminary designation of ὕλη. Ὕλη itself is established as what is cut out for, what is in fact not yet, what is still distant, ἄπειρον. There occurs a continual excluding, letting go and avoiding, and that means a relation to contraries. But all of this seems to go on without λόγος. Yet Aristotle says it should be the other way around. If so, then this concerns the relation between εἶδος and λόγος. The "representing" [*Vor-stellen*] of the εἶδος is a selecting and thus a giving notice (λόγος). The τέλος is selected out. Addressed in this way, it claims the leadership in producing; it regulates, and it does this by excluding. But is not the entire interrelation in Aristotle more simply seen inasmuch as judgment—both positive and negative judgment—pertains to ἐπιστήμη, knowledge?

But why is there this contrariness of positive and negative in λόγος? Because the essence of λόγος is notification, and because this giving notice to something is necessarily a giving of something as something. But why necessarily? Because all giving is a response to a receptive not having. This receptive taking as not having is only partially a taking into possession of something because that which is to be possessed always remains other. Partially means always in this or that respect, always as this or that. With this "as" it is always a this or that which is decided upon and separated out. But why then does the as belong to λόγος? Because notification pertains to conversance, and conversance is originally a response to exploring. *To explore*, however, is necessarily to adopt a course. It is always the choice of one way by giving up others. It is likewise the assuming of one position and the forgoing of others. This *inner boundary* belongs to conversance; the adopting of one course of exploring, and thus the simultaneous emergence of other courses which remain unexplored. This inner boundary is also the ownmost power of conversance. Therein lies the potential assurance of greatness in the venture of human existence.

But what has been said so far is only an indication of the direction of the question and the kind of problem whose resolution will clarify for us the divisive essence of λόγος. The clarification of this will likewise show how it is and why it is that λόγος, at one with its divisiveness, must be dispersed into a *multiplicity* of expository sayings and assertions; or better, why it is always already found split up and scattered in this way. The unity of conversance is always a winning back.

All λέγειν, gathering, is selecting. It is a relation to one and thereby to others, whether it be to the one *and* the others or to one *or* the other. Because λόγος is originally a selecting, it is the basic activity which guides every relation to the ἔργον as that which is selected, the τέλος. And it is only because this selection of what belongs together is gathered in the εἶδος and likewise demarcates from out of itself a material to be selected and its determinate ways, its preparation, that every producing is *gathered* in itself in terms of how it is in its ownmost meaning. Only because being gathered into one belongs to every work, no matter how unimportant and trivial, can producing a work be disseminated and careless and the work be disorderly, that is, a non-work.

The being-gathered-together of production is at play in the gathering (λέγειν) of the discussion and of the cognizance that discusses what is or is not suitable. This is that *talking to oneself* which for the most part goes on silently or as a commentary which gets lost in the work and is often seen only from outside as a bunch of disconnected words. Producing is intrinsically a talking to oneself and letting oneself talk. To tell oneself something does not just mean to form words but to want to proceed in a certain way, that is, to have already gone there in advance.

§ 14. Relationship of force and conversance

Regulative cognizance makes producing possible only if it is deployed in the exploration of individual provisions and steps which have to be carried out in a definite order to complete the production. This deployment of cognizance is what we call deliberation. It is a going over something with oneself and discussing it. This dialogue is the inner deployment of and is itself. A (silent) deliberation directs the individual steps of producing. It demands that it be the director because it is essentially an activity which has already taken into its view what is to be done and produced.

Producing is therefore in no way simply accompanied by a succession of assertions which are superimposed on it; nor is ἐπιστήμη ποιητική only a series of propositions and assertions. Rather, it is a fundamental posture toward the world, that is, toward the enclosed openness of beings. Where there is world, there is work and vice versa.

Ἐπιστήμη ποιητική is δύναμις μετὰ λόγου. This μετά does not mean an indefinite "with" in the sense of "being accompanied by" or "in addition to." This ἐπιστήμη is, in its innermost essence, μετά. This means close behind something and following it, pursuing it and led by it—by λόγος. Thus the translation: led by discourse.

We still have to answer the question (see above, p. 116) whether λόγος is understood in this passage in a *narrower* sense. Λόγος primarily means conversance and openness of what is to be brought forth. It is the outward appearance of the summons (*Vorwurf*), the εἶδος. It is also the discussion of the plan and the organization of measures taken for its execution. We can convey this in the form of assertions. The meaning of λόγος as assertion is derived from the meaning of λόγος as εἶδος (see *Met.* 27, 1032b2-3). Such an investigation pushes for a detailed discussion of something which must already be open in advance as a whole. On the other hand, this restricted meaning of λόγος is precisely the meaning which lies closest to us and the meaning which we most often encounter, and therefore also the meaning which takes over the specific role of leading in all the various ways of behaving, not only producing. The meaning of λόγος as assertion, which is restricted and derived in terms of its essential origin, is the widest in terms of its use and the range of its control. In our context, Aristotle has both meanings in mind. The inner connection of both meanings can be understood only if we establish in advance the original, essential character: conversance and openness. However, any possibility of understanding will be blocked off if we take λόγος "logically," in the current sense of the term, according to which λόγος means judgment, assertion, and the mainspring of assertion, concept.

We can only allude to the not infrequent use of λόγος in which the meaning of the word resonates out of the gospel of St. John and the Oriental gnostic teachings on wisdom, and which utterly transforms the original Greek content of the word.

In order to understand the entire context of the preceding pages, we have to emphasize again that we cannot explain and try to define the essence of λόγος through the ideas of a professor who appeals to a logic textbook for support, not even when, as in our passage, the topic is about ἀπόφασις (Θ 2, 1046b13-14). Not judgments and forms of judgment are meant here, but the inner movement and lawfulness which lies in the openness of the world and which presents itself for the Greeks primarily and essentially in λόγος and as λόγος. Only from out of all this can the fabrications first be extracted which logic and grammar then introduced as so-called forms of thought and grammatical forms. This and many other things have left us standing helpless over and against the essence of what we call language and fundamentally alienated from it. Thus on the one hand the inner neglect of language and the lack of respect for its dignity, and on the other hand the idolatry of an abstract clanging fabrication and somewhere in addition to this even a science of language, which makes its countless discoveries continually in a vacuum, without ever finding its way back to language.

§ 15. Δύναμις κατὰ κίνησιν *as capability of the striving soul*

Only if λόγος is grounded in the meaning just presented does one understand the inner connection with the whole constitution of δύναμις to which it belongs. We have seen: δύναμις μετὰ λόγου is there only where ἔμψυχον, where there is soul and the besouled in

§ 15. Δύναμις κατὰ κίνησιν *as capability*

general. But this relationship may not be accounted for simply in this way: exploring and asserting are processes of the soul, and therefore this very δύναμις μετὰ λόγου necessarily has to be a capability of the soul. But things are otherwise and in a certain sense reversed: if a δύναμις is the sort that belongs in the region of being pertaining to the soul, then not only is it directed by a λόγος, but its entire character as δύναμις is other; as δύναμις, that is, as ἀρχὴ μεταβολῆς ἐν ἄλλῳ. How such a δύναμις of the soul is constructed and how λόγος necessarily works its way into this essential structure, Aristotle attempts to show in the following sentences.

> 1046b15-22: ἐπεὶ δὲ τὰ ἐναντία οὐκ ἐγγίγνεται ἐν τῷ αὐτῷ, ἡ δ'ἐπιστήμη δύναμις τῷ λόγον ἔχειν, καὶ ἡ ψυχὴ κινήσεως ἔχει ἀρχήν, τὸ μὲν ὑγιεινὸν ὑγίειαν μόνον ποιεῖ καὶ τὸ θερμαντικὸν θερμότητα καὶ τὸ ψυκτικὸν ψυχρότητα, ὁ δ'ἐπιστήμων ἄμφω. λόγος γάρ ἐστιν ἀμφοῖν μέν, οὐχ ὁμοίως δέ, καὶ ἐν ψυχῇ ἣ ἔχει κινήσεως ἀρχήν· ὥστ'ἄμφω ἀπὸ τῆς αὐτῆς ἀρχῆς κινήσει πρὸς τὸ αὐτὸ συνάψασα.

"Since that which lies in the most extreme affinity does not get formed (at the same time) in the same being, yet since the expert understanding of something is a force on the basis of its being directed by discourse, conversance, and since the soul holds forth in itself an origin for movement, so indeed can the healthy promote merely health, that which gives warmth warmth, what cools only coolness, but in contrast, expert understanding is related to both (the contraries). For conversance is always directed at both, but not in the same way, and it belongs (according to its way of being) in a soul which itself (as such) holds forth in itself a from-out-of-which for movement. Hence it will bring both into movement, and in fact proceeding from the same origin in such a way that it brings both back together to that which is discerned as the same."

At first roughly the same theme as above, at first sight nothing new, but only a broad recapitulation that in λόγος δύναμις is related to ἐναντία as ἄμφω. And yet we ought not to read out two new essential determinations: on the one hand, it is ψυχή that is explicitly under discussion; on the other hand, it is κίνησις that is under discussion, something which is no less essential for the undecided question concerning δύναμις. More exactly, what is being discussed is the fact that ψυχή, soul, has, holds, and holds forth in itself that from out of which its self-moving occurs—ἀρχὴν κινήσεως ἔχει. And with this, λόγος and the work that it has been depicted as doing are brought together. To be sure, Aristotle gives at this point no more precise clarification of ψυχή and κίνησις and the connection of the besouled or living with the self-moving. But he speaks of it as well-known, not in itself well-known but known and clarified through what he often and in various ways and in different respects has spoken of in his lectures. In what respect he now wants the question concerning ψυχή as ἀρχὴν κινήσεως ἔχουσα to be more precisely understood makes itself known in that the mentioning of this connection occurs in a way inextricably bound to the question of λόγος and δύναμις. We too in our interpretation must give up discussing this connection extensively. Here we mention merely the investigation in which Aristotle thoroughly treats the question: *Met.* Z 7-9, *Nic. Ethics* Z, *De an.* Γ, especially chapter 9ff.

In the treatise Περὶ ψυχῆς this question is (naturally) dealt with as the most proper theme. Ψυχή is what constitutes the being of beings which have the character of living. Περὶ ψυχῆς is not a treatise on psychology but an ontology of the living overall. What lives has the fundamental character of self-moving, which does not necessarily mean changing place. Plants, too, which have their fixed location, move themselves—as growing and nourishing. The movement of the living is a self-moving, and above all for the reason that ἀεὶ ἡ κίνησις ἢ φεύγοντος ἢ διώκοντός τί ἐστιν (*De an.* Γ 9, 432b28f.), for the reason that movement always is that of fleeing or pursuing (φυγή or δίωξις). This means, however: to protect oneself from something or to take something into possession. The movement of the living is ἀεὶ ἕνεκά τινος (see b15ff.)—always for the sake of something, something which in the end is at issue, something which is the end, τέλος, what is to be accomplished—the πρακτόν. That which is at issue is necessarily an ὀρεκτόν, *something striven after* (Γ 10, 433a28).

To briefly clarify these connections: an ὀρεκτόν is something posited in a striving, through the striving as such set forth [*Vor-gestelltes*]. Striving is inherently setting-after something and as such already set-

§ 15. Δύναμις κατὰ κίνησιν *as capability*

ting-before; this comportment can, however, set aside this setting-after and is then only setting-before. Everything which we call "representing" [*Vorstellen*] and "intuiting" is inherently this "bare setting before, this bare representing"; it is not, for example, the reverse: first represented and then striven after. Thus there are manifold ways in which the ὀρεκτὸν ἢ ὀρεκτόν is manifest.

This ὀρεκτόν is, however, in each case ἀρχή: that from-out-of-which and that in reference back to which all effort is set in motion; to this effort belongs also the deliberation, and the dialogue over the right way and right means. Hence: τὸ ὀρεκτόν . . . γὰρ κινεῖ (433b11-12)—the striven after as such is what properly does the moving; it is *the* ἀρχή *of* κίνησις *that the soul has*. The soul has this ἀρχή insofar as the soul as essentially *striving*, as ὄρεξις (Γ 9, 432b7), is related to an ὀρεκτόν. The having, ἔχειν (cf. likewise λόγον ἔχον), does not simply mean: having in itself, as some sort of property, but having something in the manner of a holding-itself-in-relation-to, of a comportment—whereby that at which the comportment is directed is made known somehow in and through this comportment itself. (For this reason it is for Aristotle an important question indeed whether λόγος must not also be attributed to non-rational animals, or the beings that we name in this way.)

Where, therefore, comportment and self-moving is a production, ποίησις, and in fact an ἐπιστήμη ποιητική, a human activity, there the ἀρχή of this activity is not only and first of all the εἶδος, the λόγος, as we portrayed it earlier—the projection of what is to be produced there, the making known of the outward appearance—but at one with this it is, even indeed prior to this, already an ὀρεκτόν; as, for example, in the striving after a useful object that makes possible the holding and transporting of water. The ποιεῖν κατὰ τὴν ἐπιστήμην, production, requires "outside" the λόγος yet another that rules—ἑτέρου τινὸς κυρίου (end of chap. 9)—yet another ἀρχή, that of wanting to have at one's disposal, for example, some sort of use object. Only this needing, this wanting, leads to a producing; that is, it is the from-out-of-which for the producing as a movement—ἀρχὴ μεταβολῆς. The needing is not only the impetus, the stimulus that comes and goes, but rather the needing has in itself already its range and orientation, and with this it guides the production up to the fulfillment of the work. So there are, as it were, two ἀρχαί and yet, then again, only one (regarding νοῦς πρακτικός, see chap. 9ff.).

From this vantage point the juxtaposition of λόγος and ἀρχὴ κινήσεως in our passage (*Met.* Θ 2, 1046b15ff.) becomes comprehensible. The ἀρχή of δύναμις μετὰ λόγου is an ὀρεκτὸν πρακτὸν λεγόμενον—as something striven after, to be produced, and addressed as this or that, and as such related to the ἐναντία because ὄρεξις is necessarily δίωξις or φυγή.

And now it can finally be articulated with utmost precision what Aristotle (loc. cit.) is trying to say: he begins with the suggestion that the ἐναντία, which indeed were already thoroughly under discussion beforehand, are not present together at once in one and the same produced being. But on the other hand, it is never the case that only one of the contraries is there with the δύναμις μετὰ λόγου. Then both. Certainly. But the question is then: how? Not as something at hand, produced, but in the production and for this. In that this proceeds from (ἀπό, line 21) something which wants and is to be produced, the contrary is already co-given with this one striven-after origin, a contrary which must be avoided in this striving. The care which belongs to production *unites* precisely both in itself: holding to the right path and avoiding going off track and awry. Both what meets up with the right path and what meets up with the wrong path, both are constantly seen together, and the two are referred back together to the one out of which the whole producing is set into and held in motion, the ὀρεκτὸν πρακτόν.

In the sentences we are now discussing, Aristotle wants more to remind us of this connection rather than to expressly elaborate on it. We saw how and in what sense δύναμις μετὰ λόγου is indissociable from something besouled (ἔμψυχον). At the same time we saw that here soul is in no way a thing that acts and makes itself felt in a body; rather, we saw that Aristotle exposes a very definite fundamental structure of living beings and incorporates into this structure itself the enactment of λέγειν and λόγος as an occurrence belonging to it.

§ 16. The inner divisiveness and finitude of δύναμις μετὰ λόγου

We maintain: the inner divisiveness of λόγος is the origin and root of the proliferation into individual λόγοι; this occurs for the most part not as forged interpretations but in the "speaking to oneself" of producing. The gatheredness of producing springs forth out of the essence of λόγος as gathering. It is from this perspective that the μετὰ (λόγου) is to be understood. In all this it is important not to conceive of λόγος in terms of "logic," but to proceed in the opposite direction. —Now the δύναμις μετὰ λόγου as κίνησις enters into the discussion. How is it related to κίνησις? The movement here is that of the living, the ψυχή; this has the ἀρχή of movement—ἀρχὴν κινήσεως ἔχει. The movement of the soul consists in striving, ὄρεξις, and is either flight or pursuit. What is striven after, the ὀρεκτόν, is not itself a mere object that is represented but the one that moves; it is this as λεγόμενον εἶδος. (The same fundamental connection of κίνησις and ὀρεκτόν is found in the concluding book of the *Physics*. That which in the primary sense does the moving moves ὡς ἐρώμενον [see *Met*. Λ 7, 1072b3]; ἔρως is characteristic of a specifically Platonic way of seeing the living kind of movement, which recurs in Aristotle in a modified form.) The ἀρχή, that from out of which everything living is set into motion, is thus had and held by the soul (ἀρχή ἐχομένη), and in fact in various relations, to be delineated through the phenomena εἶδος, τέλος, ὀρεκτόν, and λόγος. These define one and the same ἀρχή back upon which the whole occurrence and inner constitution of δύναμις is referred.

Every production of something, in general every δύναμις μετὰ λόγου, prepares for itself, and this necessarily, through its proper way of proceeding, the continually concomitant opportunity for mistaking, neglecting, overlooking, and failing; thus every force carries *in itself* and *for itself* the possibility of sinking into *un-force*. This negativum does not simply stand beside the positive of force as its opposite but haunts this force in the force itself, and this because every force of this type according to its essence is invested with divisiveness and so with a "not." Yet for Aristotle and antiquity there was almost no essential urge to pursue these questions, to the extent that for the Greeks it was the question about being which above all and first of all had to render a comprehension of what is questioned.

Met. Θ 2, 1046b22-24, suggests how the clarification of δύναμις μετὰ λόγου and ἄνευ λόγου applies to the concept of δυνατόν:

διὸ τὰ κατὰ λόγον δυνατὰ τοῖς ἄνευ λόγου δυνατοῖς ποιεῖ τἀναντία· μιᾷ γὰρ ἀρχῇ περιέχεται, τῷ λόγῳ.

"Hence the forces according to discourse (the capable) enact the contraries for the forces without discourse; for (the capable) is encompassed by one (single) origin; this one is conversance."

As it stands, the sentence is unclear and therefore ambiguous. The capable that is directed by discourse does the opposite to what is without discourse. This primarily means: it is not directed at something singular, as is what is without discourse, but, in contrast, it is directed at a one and its other, and that is to say, at contraries. Because the capability that is directed by discourse is related to contraries from the bottom up, therefore it is as a whole inherently at the same time contrary to capability that is without discourse. In the ποιεῖν τἀναντία both are pulled together: both the relating itself to contraries and thus, in contrast to the δύναμις ἄνευ λόγου, the comporting itself contrarily. In terms of content, the sentence does not offer anything new. Only the formulation is noteworthy, the way in which the λόγος is named the unitary, singular ἀρχή; here it is indeed corroborated that the λόγος is no concomitant phenomenon in ποίησις and ἐπιστήμη ποιητική but constitutes the innermost framework. In this the μετά is properly determined.

The concluding sentence of the whole chapter also offers a thought that has already been touched upon; it is not immediately apparent what the addition of this sentence at this point is supposed to do.

1046b24-28: φανερὸν δὲ καὶ ὅτι τῇ μὲν τοῦ εὖ δυνάμει ἀκολουθεῖ ἡ

§ 16. Inner divisiveness and finitude 133

τοῦ μόνον ποιῆσαι ἢ παθεῖν δύναμις, ταύτῃ δ' ἐκείνη οὐκ αἰεί· ἀνάγκη γὰρ τὸν εὖ ποιοῦ· τα καὶ ποιεῖν, τὸν δὲ μόνον ποιοῦντα οὐκ ἀνάγκη καὶ εὖ ποιεῖν.

"It is then also manifest that being forceful in the right way is followed after by force, the force simply to do something or bear something, but being forceful in the right way does not always follow after force; for the one who produces in the right way must also necessarily (first of all) produce, namely the one who simply produces but not necessarily also in the right way."

The thought is clear: The force for producing in the right way presupposes that in general a force for producing is there; but, conversely, the latter does not already imply the former. Thus it is plain in what sense Aristotle here understands the expression ἀκολουθεῖν. Being-capable-at-all of something "follows" being-capable-in-the-right-way. We would say the reverse. But here ἀκολουθεῖν means "to follow" in the sense of "constantly going after," "always already going along with something"; if viewed in terms of that which is followed after, this means: this latter, which is constantly followed after by something, carries with itself and along with itself this something which follows after it. And indeed this is so in the very definite sense that this something which always already goes along with is the condition of possibility for that with which it goes along, which it follows. It is important that we are clear on this. In this expression of following we have, understood in a Greek way, the formulation of the relationship which we learned to express as the connection of *a priori* conditions. Following means here: to go in advance, not only to come afterwards. What πρότερον φύσει is—the earlier in terms of the matter, this itself has nothing more behind it, but to be sure it always stands behind that which it conditions in terms of the matter and in this way goes after it.

But just there where the ἀκολουθεῖν means "to follow after" in the sense of following right after, there the meaning is not a coming-later in a temporal sense, yet neither is it the so-called logical succession, but rather the essential being-conditioned. This ἀκολουθεῖν plays a great role in Aristotle in, for example, his doctrine on the essence of time. Here there arises a succession and structural relationship between μέγεθος, κίνησις, and χρόνος, extension, movement, and time. Here the ἀκολουθεῖν in fact is meant in the reverse direction as "following upon"; in no way in the sense of a logical succession or even an actual emergence, but in relation to the order of the basic structure of the essence of time; this is grounded in movement, and movement in turn is grounded in extension in general. Whether in this way or that way, ἀκολουθεῖν is used in the sense of essential belongingness; cf. *Met.* A 1, 981a24ff.

The above sentence is thus transparent in its content; but precisely because this is the case, we ask ourselves why it is there. It has, in fact, nothing to do with the guiding question of the whole chapter. And yet—if this connection of the εὖ (in the right way) and the ποιεῖν was already hinted at earlier, and so was only a clue, not yet a genuine grounding—then only now, precisely on the basis of the discussion of λόγος and its belonging to δύναμις, do we first grasp that within which the εὖ that belongs to δύναμις has its roots. Why does there belong to a force the "in the right way"—and this means "in the not right way," "in an indifferent way"? Why does there belong to a force necessarily the "in each case such and such," in general: the *how?* Because force as δύναμις μετὰ λόγου is from the bottom up *doubly* directed and *bi*furcated. And because, then, the force which is directed by discourse is in an original sense *of the not*, that is, shot through with this not and no, for this reason the how is not only altogether essentially necessary but consequently always decisive. For such a force, that is, for such a capability, the how belongs in the governing realm of that of which the force is capable. The how is not a concomitant property but that which is co-decided in the capability and with it.

With δύναμις without discourse the situation is otherwise; true, we also speak here of an εὖ, that something which warms gives off good heat or bad. But the "good" and "bad" belong to this force in an entirely different way than is the case with δύναμις directed by discourse. To be sure, these are connections which no longer lie in the realm of Aristotelian questions.

But that Aristotle wants to steer our vision to the inner connection between the how of a force and the divisiveness implicit in it, just this

§ 16. Inner divisiveness and finitude

is stated by the φανερὸν δὲ καί (and so it is then also manifest; *Met.* Θ 2, 1046b24). The divisiveness is not lacking in the δύναμις ἄλογος; it is simply an entirely different one in accordance with the essence of this force. Insofar as every force is directed at a singular—μία ἑνός (b6)—it is excluded from all else in such a way that it is precisely not conversant with the other contrary and does not have it in its realm of mastery. But an exclusion nevertheless determines the singularity of that for which it is forceful. And this exclusion offers again a clue for the inner essential belonging of withdrawal and notness [*Nichtigkeit*] to the essence of force.

Over and above the individual discussions, however, this is the decisive content of the second chapter, the fact that therein the essential notness, that is to say, the inner *finitude* of every force as such, is illuminated. With this is not meant the thrusting up against external boundaries and constraints and advancing no further, nor the simple eventual failing; rather, the inner essential finitude of every δύναμις lies in the decision over this way or that required from out of itself and indissociable from its enactment. Where there is force and power, there is finitude. Hence God is not powerful, and "omnipotence," considered properly, is a concept which dissolves, like all its companions, into thin air and is unthinkable. Or, if God is powerful, then he is finite and in any case something other than what is thought in the vulgar representation of a God who can do anything and thus is degraded to an omnipresent being.

The interpretation of the second chapter—gathered up together—leads us far afield from the picture which originally offered itself, according to which what was at issue was a mere division of δυνάμεις. By now it has become clear: what is at issue is the elucidation of the essence of δύναμις κατὰ κίνησιν in general and the response to the question: what is this δύναμις, what makes up its what-being? The first two chapters come together in the end in a unified and unambiguous inquiry. And it may be profitable to go back through both chapters one more time and pull them together; this remains for each of us to do.

We are torn away from this concentrated fathoming of the essence of δύναμις κατὰ κίνησιν by the beginning of the subsequent third chapter and by the chapter itself. At the same time, we run up against the first great obstacle which, after this enticing beginning, places itself in the way of pursuing a unitary structure for the entire treatise. The decisions about whether or not portions of text belong or do not belong to the structure of the treatise, about their appropriateness or inappropriateness, these decisions depend, as usual, upon the level of comprehension of the matter attained in each case. And that goes especially for the following chapter, which everyone heretofore has taken much too lightly, as they had to, because indeed the entire treatise of which it is a part has long been vulgarized through the use of worn-out catchwords, and because our sense for the questions treated therein has become dulled.

Chapter Three

Metaphysics Θ 3. The Actuality of Δύναμις κατὰ Κίνησιν or Capability

§ 17. The position and theme of this chapter and its connection to the thesis of the Megarians

Let us begin by briefly recalling the general outline of the entire treatise. Both phenomena, δύναμις and ἐνέργεια, are to be discussed first according to their ordinary meaning, so as to make the transition to a treatment of δύναμις and ἐνέργεια ἐπὶ πλέον according to their proper philosophical meaning (cf. p. 40 above). In addition to the preliminary sketch of this very general outline found at the beginning of the first chapter, we also find at the beginning of the sixth chapter a still more extensive and explicit remark concerning the structure of the treatise. Here it is stated: ἐπεὶ δὲ περὶ τῆς κατὰ κίνησιν λεγομένης δυνάμεως εἴρηται—"since δύναμις κατὰ κίνησιν has now been dealt with..." From this we are to understand that chapters one through five deal with δύναμις κατὰ κίνησιν. Only a cursory investigation already shows, then, that chapter six begins the treatment of ἐνέργεια, and precisely ἡ ἐνέργεια ἐπὶ πλέον. Ἐνέργεια κατὰ κίνησιν as well as δύναμις ἐπὶ πλέον, however, do not enter into the discussion at all.

So it appears at first. But we still have the remarkable fact that, viewed in this manner, absolutely no transition takes place either from δύναμις κατὰ κίνησιν to δύναμις ἐπὶ πλέον or, correspondingly, from ἐνέργεια κατὰ κίνησιν to ἐνέργεια ἐπὶ πλέον. We proceed directly from δύναμις to ἐνέργεια. But what is this supposed to mean, that the discussion of ἐνέργεια ἐπὶ πλέον follows the discussion of δύναμις κατὰ κίνησιν? This is an impossible leap. Until now we have by no means brought the discussion of δύναμις κατὰ κίνησιν to its completion. But in this subsequent third chapter, the topic does suddenly turn toward ἐνέργεια.

Is a bridge thereby formed from δύναμις κατὰ κίνησιν to ἐνέργεια? And yet—according to the explicit remark at the beginning of chapter six—it must be taken as a serious violation against the structure of the whole treatise that prior to the sixth chapter the topic turns to ἐνέργεια at all. Thus the position of the third chapter, as well as that of the fourth chapter, becomes thoroughly ambiguous.

Then comes still another surprise. The beginning of the third chapter reveals that Aristotle suddenly becomes involved in a confrontation with polemical questions. Those who are familiar with Aristotle know that it is a characteristic practice of his to introduce a question first by means of a confrontation with other, earlier opinions. But this hardly amounts to an arbitrary critique and rejection for the sake of placing his own standpoint in the correct light. These confrontations develop instead what is at issue in the question and set forth the extent to which previous attempts have followed paths which were dead ends: these are paths that do not provide a way out into the open—ἀπορίαι. These discussions of the aporia already purposely exhibit the possible content of the question within certain limits (ἀπορία—διαπορεῖν—εὐπορία). They are not simply negatively critical polemics, and neither are they the detached concern of a so-called aloof "aporetic," who fanatically offers only conflicts and antinomies, and wants to let these antinomies stand, and even to hypostatize and inflate them merely for the sake of argument. The aporia point only toward the lack of originality in the posing of the question—that is, they provide the impetus toward the necessary repetition of the question.

In this case, however, the inquiry itself is by no means ushered in through such confrontations. Instead these do not emerge until chapter three, where a definite conception of the Megarians with respect to δύναμις comes into the discussion. And what would prevent Aristotle from following the opposite procedure just once, letting the aporia come after the proper thematic inquiry or even letting them be situated in its very midst? Certainly this is entirely possible. And yet before we come to a decision about this or about the further movement of the inquiry in general and make something of the adequacy of its

§ 17. Thesis of the Megarians

general articulation, let us for once attempt very simply to make clear what is at issue in this debate with the Megarians.

1046b29-30: εἰσὶ δέ τινες οἵ φασιν, οἷον οἱ Μεγαρικοί, ὅταν ἐνεργῇ μόνον δύνασθαι, ὅταν δὲ μὴ ἐνεργῇ οὐ δύνασθαι, οἷον τὸν μὴ οἰκοδομοῦντα οὐ δύνασθαι οἰκοδομεῖν, ἀλλὰ τὸν οἰκοδομοῦντα ὅταν οἰκοδομῇ· ὁμοίως δὲ καὶ ἐπὶ τῶν ἄλλων. οἷς τὰ συμβαίνοντα ἄτοπα οὐ καλεπὸν ἰδεῖν.

"There are, however, certain people, such as the Megarians, who say that the ability to do something is present only while a force is at work, but when it is not at work, then there is no such ability. For example, a builder who is not building is not able to build, unlike the builder who is building. The same could be said of other kinds of force. It is not difficult to see that what is proposed by this statement cannot in any way be accommodated."

By introducing this Megarian thesis, Aristotle furthers his inquiry. Who are these Megarians? An answer must be drawn precisely from this Aristotelian passage itself, since none of their writings have been handed down to us. Occasional fragmentary doctrines and statements of theirs are mentioned in the writings of the Stoics, in Sextus Empiricus, Alexander of Aphrodisias, and Simplicius. They make up a philosophical orientation and school which, like that of Plato, had its origins in Socrates; their founder is Euclid of Megara (not the mathematician). They attempted to bring together the philosophical activity of Socrates and the teaching of the Eleatics, Parmenides and Zeno. The confrontation with these thinkers certainly pertained also to Platonic as well as Aristotelian philosophy, both being contemporaneous with the Megarians. One of the questions, or even the central question, of all three orientations concerned the essence and possibility of movement. And this means in a certain sense the question of the being of that which is not, or in other words, the question of the essence of the not and of being in general. The fact that the Megarians troubled themselves with this question, and that Aristotle concerned himself with them in such a prominent passage, as did Plato in his *Sophist* (246bff.), shows that they were not spurious verbalists who sought to procure a position by means of claptrap and empty sophisms. This is the usual portrayal, derived mostly from reports about the later period of their school, whereby one also forgets to mention that the later students of the schools of Plato and Aristotle were not much more noteworthy than the late Megarians. It would seem, then, that these contemporaries of Plato and Aristotle were of the same rank, although it was their fate to have been forgotten in history.

*

Last time we brought to a close the interpretation of the second chapter. This resulted in our being directed toward a double determination: (1) Every δύναμις is prepared at all times for the occasion of lapse and failure; the possibility of sinking into unforce is inherent in it; this does not merely relate to it circumstantially, as though unforce were something other. (2) The conditioned relationship of ποιεῖν and εὖ ποιεῖν is such that the former follows the latter (not the reverse); ἀκολοθεῖν means to go after, always already to stand in the background; it is a matter of that which is never to be circumvented. And while it is also employed in the reverse order (compare Aristotle's treatise on time), there it always refers to the structural relationship of essential conditions to what is conditioned, and thus is primarily not a logically reducible order and certainly not a temporal sequence. The origin of the necessity of the how of a capability lies in the inner and essential divisiveness of that capability itself. —The theme for both chapters one and two is the essence of δύναμις, that is, the determination of what it is. Now with the third chapter comes the first great disruption, at least according to the conventional conception of the interpreter and what initially presents itself. And it is in fact remarkable that Aristotle in the middle of his discussion starts a critical confrontation with others. The question is whether this confrontation primarily follows chapters one and two or whether it does not perhaps anticipate and prefigure something else. It was Aristotle's custom to begin a discussion with an aporia, which for him had a positive meaning. This has to do not with a mere undecidability of the question but rather with getting onto the right path (διαπορεῖν) and heading toward an εὐπορία. A Megarian thesis is posed as the

§ 17. Thesis of the Megarians

consideration begins. Who are these Megarians? No writings have survived. They were Socratics and "Eleatic," contemporaries of Plato and Aristotle. Their principal question concerned the essence and possibility of movement.

That different conceptions in relation to the question of the essence and possibility of movement still could and even had to stand over and against Plato and Aristotle is easy to see, if one has even a slight appreciation for the tremendous struggle which these two thinkers—Plato and Aristotle—had to take upon themselves if they were somehow to gain even a tenuous footing within the dark and precarious realm of this question. The question concerning what and how movement "is" posed at that time, as it fundamentally still does today, its own peculiar difficulty not only in the enigmatic essence of movement as such, but in the interpretation and understanding of being, in the light of which one first becomes troubled by the being of movement.

The Megarians denied the possibility of the actuality of movement, according to the fundamental Eleatic principle of the being, wherein only the being is and the non-being is not. And yet every being that is in some manner tainted and pervaded by the nothing is non-being—thus the not-yet-being as well as the no-longer-being. What is in movement, however, suddenly changes, moves out of one thing into another, is no longer that but not yet this. What moves is in this sense non-being from "two sides": it is respectively not yet what it will be and no longer what it was. Being [*Seiend*] is only what is present and at hand.

Now we have to assume that the Megarians did not simply rehash the old theses, but rather sought to defend the Eleatic theses in this confrontation with Plato and Aristotle and their doctrines of movement. And not only this—and what I am about to say is for you perhaps an empty assertion at present, but for me it is a personal conviction—but one might rightfully doubt whether Plato and Aristotle actually comprehended and overcame the central objections of the Megarians. With this it may also remain undecided whether the Megarians themselves knew what they for their part fundamentally wanted. No true and great wanting knows, in a manner which can actually be stated conceptually, what it has wanted. To discern this is the business of those who come later. But this subsequent improved understanding permits no superiority. The following interpretation should show the extent to which this doubt is justified and must be affirmed (as to whether Plato and Aristotle actually overcame the Megarian objections). Even if the refutation of the Megarians presents no difficulty, as Aristotle states in our passage, there still remains the question whether Aristotle grasped or even ever could have grasped the ultimate difficulty—and this means the primary difficulty—through which the Megarian argument took on its full weight and validity.

Let us inquire now in a more definite manner, and in relation to our treatise. What is the connection between the Megarian doctrine and the Aristotelian thematic, which concerns δύναμις as δύναμις κατὰ κίνησιν? This δύναμις holds an essential reference to κίνησις, insofar as δύναμις is what it is: ἀρχὴ μεταβολῆς. Now if μεταβολή is in itself essentially impossible, then this also pertains first and foremost to that which is claimed as ἀρχὴ μεταβολῆς. An origin for that which in itself cannot be, is itself senseless. If δύναμις is supposed to be such an ἀρχή, then it cannot be at all. With this the initial strangeness of the relation between chapter three and the earlier chapters already vanishes. The decisive question simply is: Does the Aristotelian discussion of the Megarians seek only a polemical clarification and a supplemental determination of the exposition in chapters one and two, or does chapter three issue in a new question? Does the treatise proceed along its path in this manner? If so, in what sense?

One can easily see how exceptionally important it is that we first attain clarity with regard to the questioning and results of chapters one and two. And yet the usual way in which these two chapters have been speciously interpreted and popularized has thus far prevented the achievement of an adequate preparation for an understanding of what follows. As a result, these interpretations, if one may even call them that, only heighten the confusion of the questioning, a confusion which, even in Aristotle's time as the debate took place, did not exist accidentally but was and still is rooted in the immeasurable difficulty of the matter in question. Accordingly, extreme care and rigor are demanded precisely in the unfolding of this undecided question.

In order not to strengthen the suspicion that we are trying to force a connection between chapter three and chapters one and two, let us now set aside the possible connection which we have just suggested.

§ 17. Thesis of the Megarians

Instead, we are now going to attempt much more, to unfold the whole treatise purely in the context of chapter three. We proceeded as though we knew nothing of the Megarians, and yet in fact we need not make this assumption. We can let Aristotle tell us something about them.

The text which we translated yielded one thing, that the Megarians already responded to the Aristotelian doctrine concerning δύναμις, and even disputed a very definite point. With this dispute, however, they are touching upon a central problem. (This, again, is primarily our assertion.) There is, then, still something to which we have to direct ourselves: Aristotle refers to the thesis of the Megarians only, but not to what for them evidently grounded this thesis.

What is at issue? Δύναμις κατὰ κίνησιν. As an example, an ἐπιστήμη ποιητική is mentioned, namely the οἰκοδομική, the art of building. Thus, in the terminology that we established earlier, what is at issue is a capability. More specifically, this deals with the question of *whether, when,* and *how* such a *capability could actually be present*, precisely as the capability that it is. The *Megarian thesis* states (1046b29/30): ὅταν ἐνεργῇ μόνον δύνασθαι, that is, δύναμιν ὑπάρχειν. I translate this and in doing so stress its decisive meaning: "While force is at work, only then is the ability to do something present." What is forming and guiding our understanding and interpretation of this entire chapter is the translation of ἐνεργῇ. Ἐνεργεῖν means to be at work (and not simply to be actual). If a capability for something is "at work"—and this means engaged in the production of that for which it is a capability—then we say concisely that something "actualizes" itself which formerly was only something potential.

What is at issue here is thus the question concerning the actuality of the potential, that is, ἐνέργεια, which also then is indeed the recurrent topic of this third chapter. But ἐνέργεια is not supposed to be dealt with until after chapter five. Thus Aristotle already corrupts the structure of his own treatise. An observer as careful as Bonitz also noted this. *Ad definiendam* δύναμιν *iam hoc loco adhiber notionem* ἐνεργείας *de qua infra demum, inde a cap. 6, uberius disputabit.*[1] Thus the question is (1) whether the topic here is the definition of δύναμις,

---

1. *Commentarius*, p. 387.

or not rather something else, and (2) whether, in addition, the concept of ἐνέργεια ἐπὶ πλέον is not already being brought into play. It does not suffice simply to establish extrinsically the occurrence of ἐνέργεια and to identify this immediately with the later formulation, when to the contrary it is quite clearly emphasized at the very beginning of the treatise that the later philosophical concept must first be achieved.

Certainly in the very first sentence of chapter three ἐνεργεῖν makes an appearance, and we must take heed of this, but the question simply is: Is this the ἐνέργεια of chapter six, namely ἐνέργεια ἐπὶ πλέον, actuality in contrast to potentiality? Or is this not ultimately what has always been overlooked by the interpreters, the ἐνέργεια κατὰ κίνησιν, the being at work as distinguished from . . . Yes, and now, distinguished from what? From capability, from mere capability? Exactly! And just this, whether and how such a "mere capability" can "be," that is the question. Only in the form of this question can the ἐνέργεια κατὰ κίνησιν enter the discussion at all. Why this is so, and why Aristotle did not devote a specific section to this and did not write a transitional passage, is deeply rooted in the matter. Neither is it accidental, on the other hand, that those who have offered explanations have confused the whole problem, but this too is rooted in the obscurity of the matter. It is therefore incumbent upon us first of all to discover the true location of the basic difficulty and to free up and expose this question itself from its various sides.

The Megarian thesis cited by Aristotle states: a δύναμις is only when it actualizes itself. This thesis pertains not so much to the what-being of δύναμις, what one commonly calls the essence, the *essentia*, what is commonly grasped in the "definition," but rather it pertains to the how of its being present, the *existentia*. How is that which has the essence of what was presented in chapters one and two actually present when it actually is? The Megarians looked for this being present of a capability in the actualization—that is, in the *enactment* of the capability. If the capability is not engaged in enactment, then it simply does not exist. Mere capabilities which are not enacted do not lack only factical existence, they cannot be at all. From this Megarian thesis we must at first assume that in the Aristotelian doctrine the question concerning how a δύναμις qua δύναμις is present

§ 17. Thesis of the Megarians 145

was either erroneously or inadequately answered or simply not answered at all—that indeed the question was not posed even once. On the other hand, it is clear that if the question concerning the manner of being present of a δύναμις qua δύναμις is posed, then it can be answered only with a view to the essence of δύναμις itself. How something is, is essentially co-determined by what it is. I deliberately say *co*-determined by the essence, that is, not by this alone. In any case, the how of being present does not allow itself to be simply deduced from the essence for the simple reason that the essence for its part is comprehensible only by passing through something present of this essence—or, as we say, through something "thought of" as present.

All of this indicates only that an answer to the question concerning the how of the being present of a capability as capability cannot be achieved without a view toward that essence which was developed in the earlier chapters.

And so at last we must more resolutely ask: How does this question about the being present of forces become so entangling that it precipitates these laborious discussions? There are certainly countless capabilities; we come across them continually and inconspicuously. We know the shoemaker, the baker, the potter, and the cabinetmaker; with them there are certainly very definite capabilities present. The potter, for example, is the one sitting in the tavern. He is the one who can make mugs; he is the one capable of producing them. With him a capability is actually there. Good, but *how* then? Where and how then is his capability? He does not carry it with him in his pocket, like his pocketknife, for then the question would be easily resolved. Neither is his capability to be found in the anatomical structure of his body. Perhaps his hands have a unique contour, but this is at best a consequence of the fact that the capability is in him, and not of the capability itself. We say that the capability is "in him." Where within? In his brain? There we would search in vain. In the soul, of course! But what does "soul" mean? And how then is the soul present?

What initially seemed so obvious, that the capability for making mugs is at hand with the potter sitting in the tavern, has now become completely abstruse. But we will not yet let ourselves be persuaded that the capability is not there, since otherwise even the potter would not be there; or does he leave this capability at home when he goes to have a beer? It should now be more clear already that the Megarians find a question here and want to see it raised. To be sure, the answer which they give to the question concerning the manner of being present of a capability is by no means to be brushed aside. In any case, this answer shows that they seriously pursued the question: A capability is in every case present, that is, actual, when it is actualized. And it is actualized when it enacts what it is capable of. If first of all it is capable only of something which actually can be, then capability is precisely only the potential of something actual. The potential is, however, what is not yet present. A capability can be called present, and thus being [*seiend*], only if it is engaged in its enactment. Only the builder who is building has capability.

In this deliberation the true and false become confused. It is incontestable that a difference persists between a capability which is merely acquired through practice and one which is in fact employed, and that this difference somehow pertains to the being of a capability. It is further incontestable that the capability which is merely a being-practiced-in represents something like the "potential" over and against what is practiced as something actual and actualized. And yet it could just as rightfully be said that the capability which is not practiced is not only something potential, but also indeed already something present. A potential capability is something other than an actual capability; an actual capability, however, does not need to be engaged in its enactment and actualization in order to be as such.

With this the decisive question has already become more precisely determined: *How "is" a capability, thought of not only as potential but rather as actually present, although not being actualized?* Does Aristotle give an answer to this question? How did he confront the thesis of the Megarians? Does this thesis give rise to a positive solution to the question, or at least its elaboration? Does the clarification of the essence of δύναμις κατὰ κίνησιν make any progress here? How is the progression of the inquiry to be understood accordingly? Does Aristotle remain on the path upon which he started out, or does the whole structure of the treatise shatter after the second chapter, as the

§ 17. Thesis of the Megarians 147

interpreters contend? The interpretation of the text has to answer these questions. To be sure, only with these questions do we gain even a glimpse into what Aristotle wanted to say.

But first it is necessary to add another and entirely decisive difficulty to all the ones which thus far have been developed and which lie in the matter itself. A capability is, according to the Megarian thesis, only when it is engaged in enactment, ὅταν ἐνεργεῖ—when it is at work. To be at work means to be busied with producing, to be immersed in it; here, however, the work itself is not yet a work; it is this only once completed. Once completed, however, the producing for its part has become superfluous and no longer is. The being present of a capability is the producing itself as ποίησις—that is, as κίνησις. But if the Megarians see here, in the movement of producing, the actuality of capability, then this contradicts their basic conception of being and actuality, which supposedly is Eleatic, and according to which movement simply is not and never is: it is non-being. But if this absurdity cannot be attributed to the Megarians, that precisely where the actuality of something (namely of δύναμις) is at issue, they consider it to be nothing other than what they at bottom call the non-actual, then the traditional conception of the historians of philosophy in relation to the basic character of the Megarians as Eleatic is untenable.

Yet there is still another possibility, namely that the Megarians did indeed want to find the actuality of capability in its enactment but, precisely because of their Eleatic orientation, they were not in a position to comprehend the essence of enactment. They were much more inclined to misinterpret it.

If, however, the manner and mode of being present of δύναμις involves at all the being moved of its enactment, then it is precisely Aristotle who first decisively dealt with this kind of being, insofar as he undertook to clarify the essence of movement for the very purpose of determining and making visible movement and being moved as a proper mode of being actual. If, nevertheless, the manner of being present of δύναμις does not lie in being moved, in the enactment of producing, as the Megarians wanted, and if, on the other hand, the enactment of capability is again not without a relation to the essence

148   *Metaphysics* Θ 3.

of capability, then being moved as a kind of actuality could offer at the very least a guiding thread in posing the question concerning the unenacted but no less actual capability.

With all these deliberations we have generally circumscribed what is required for achieving a philosophical understanding of chapter three.

*

§ 18. *The beginning of Aristotle's confrontation with the Megarians*

a) Is the actuality of capability to be found in having or in its enactment?

Aristotle begins chapter three with a critical confrontation directed against the Megarians. Their main concern is the question concerning the possible actuality of movement, a possibility which they deny. Behind this is concealed a fundamental question of philosophy. What connection does the Megarian doctrine have with the Aristotelian theme? This theme in general is δύναμις κατὰ κίνησιν or, more exactly, its essence as it was developed in chapters one and two. If the Megarians deny the possibility of movement, this does away with the essence of δύναμις κατὰ κίνησιν. Thus chapter three would serve only to reinforce the determination of this essence. And yet is this what is at issue here? Already the first sentence tells us something else. It does not address the question of the "what" but rather the "how" of being present. And where is this located? According to the Megarians, in ἐνεργεῖν. Thus the theme is ἐνέργεια. But how? Not as the disturbance of the structure of the treatise but as the furtherance of the problem. Accordingly, what is at stake in chapter three is no longer a definition but rather ἐνέργεια. But this is not yet the ἐνέργεια which begins with chapter six, but rather ἐνέργεια κατὰ κίνησιν. What does the Megarian thesis now signify in relation to this theme? The actuality of δύναμις is seen in its enactment. Here lies the "ac-

§ 18. Beginning of Aristotle's confrontation

tualization" of a capability; otherwise it is only "capable," "in potentiality." The potential, however, is not yet the actual. But is an unenacted capability merely something potential? Or is it already also actual, even though not enacted? Thus the question concerning the actuality of capability becomes more sharply defined. And yet, on the other hand, if it is now to be assumed that they deny movement, the thesis of the Megarians becomes incomprehensible. (But is enactment perhaps something different?)

Now let us look at how Aristotle encounters the Megarian thesis. We already know one thing, that he does not discuss the thesis itself but instead asks about what follows if this thesis is assumed. The consequences of this assumption, however, are utterly unacceptable and untenable. Therefore the thesis itself must be dismissed as untenable. Aristotle even emphasizes the ease with which through such a demonstration the impossibility of the Megarian thesis becomes evident.

1046b33-36: δῆλον γὰρ ὅτι οὔτ' οἰκοδόμος ἔσται ἐὰν μὴ οἰκοδομῇ. τὸ γὰρ οἰκοδόμῳ εἶναι τὸ δυνατῷ εἶναί ἐστιν οἰκοδομεῖν· ὁμοίως δὲ καὶ ἐπὶ τῶν ἄλλων τεχνῶν.

"For it is clear (from the presupposition of the Megarian thesis) also that no builder could be if he is not building, since being a builder means being capable of building. This is equally the case for the other kinds of production."

According to the Megarian thesis, builders exist, insofar as they are builders, only if they are engaged in the act of building. To make the consequences of this clear, it would thus be completely impossible to commission a builder to build a house, since he is in fact no builder at all if he is not yet building. To this Aristotle rejoins that being a builder means first of all being capable of building. But is this an answer to the question found in the Megarian thesis? Perhaps being a builder does have its essence in being capable of building, but with this it is not yet determined in what manner such a potential capability is actual. Precisely because this actuality can be nothing other than actualization in enactment do the Megarians especially deny Aristotle's essential determination. And so the two interpretations stand in sharp contrast to one another: *On the basis of his essential determi-*

nation of δύναμις, *Aristotle takes the question concerning the being present of* δύναμις *to be ostensibly decided; the Megarians deny this essence fundamentally because the being present of a* δύναμις *is only potential.*

This conflict obviously shall not be resolved by means of a formal deliberation. It requires a renewed approach to the matter being dealt with. Does this happen and in what way? Does Aristotle simply want to demonstrate to the Megarians the absurdity of what follows from their thesis, or is he after something else?

In order to give a preliminary and general answer to this question, we can say not only that Aristotle in fact nowhere engages in a merely formal refutation of the Megarians, but that even where it seems as though his refutation is reducible to a cheap quibbling over consistency, even there he is pushing for an elaboration of very definite phenomena. Not only that, but this whole critical confrontation is nothing other than the preparation for a positive clarification of δυνατὸν εἶναι in regard to its manner of εἶναι (in the sense of actuality). And in this manner the inquiry again takes an essential step forward. This further advancement of the question, however, is no mere deviation from the organization of the treatise, as though it might be an uncalculated presumption of what is to be developed only later in chapter six. Instead, here in this chapter we come across the genuine preparation for and the grounding of the transition from δύναμις καὶ ἐνέργεια κατὰ κίνησιν to ἐνέργεια καὶ δύναμις ἐπὶ πλέον.

Accordingly, we divide the whole thematic discussion into two phases. 1046b36-1047a20: the preparatory critique of the Megarian thesis on the basis of various arguments. 1047a20-b2: the positive thematization and determination of δυνατὸν εἶναι and of ἐνέργεια κατὰ κίνησιν (p. 184ff.). We shall in turn further divide these two phases into individual sections for the purposes of highlighting the progression of thought.

1046b36-1047a4. εἰ οὖν ἀδύνατον τὰς τοιαύτας ἔχειν τέχνας μὴ μανθάνοντά ποτε καὶ λαβόντα, καὶ μὴ ἔχειν μὴ ἀποβαλόντα ποτέ (ἢ γὰρ λήθῃ ἢ πάθει τινὶ ἢ χρόνῳ· οὐ γὰρ δὴ τοῦ γε πράγματος φθαρέντος, εἰ [instead of ἀεί, H.] γάρ ἐστιν), ὅταν παύσηται, οὐχ ἕξει τὴν τέχνην, πάλιν δ' εὐθὺς οἰκοδομήσει πῶς λαβών.

§ 18. Beginning of Aristotle's confrontation 151

"If it is impossible then to possess these kinds of expertise in something without having learned and acquired them once (before), and if it is just as impossible to cease possessing such kinds of expertise, unless one has already once given them up, which can happen either through forgetting, mishap or through time, then certainly it is not due to this that what producing in each case has to do with is destroyed, for if this were so . . . ," thus if the possession or dispossession of δύναμις is in a similar fashion impossible without the occurrence of what has been stated, "how should this expert cease in having this expertise in producing when he (merely) stops (producing)? And, vice versa, how is he supposed to have reacquired this expertise if he should suddenly resume building?"

We now want to scrutinize through a series of statements, which are directly evident if considered according to Aristotle's manner of questioning, the challenge that Aristotle returns to the Megarians which here takes the form of a hypothetical question. The preceding alteration in the text—from ἀεί to εἰ—and the translation of πρᾶγμα must be justified within their relevant contexts.

The ἔχειν of a τέχνη is bound to a previous learning and acquisition; no longer possessing, μὴ ἔχειν, is bound to a giving up. If this is so, then it is also clear that merely ceasing to enact a τέχνη in no way needs to signify already no longer having it. And vice versa, the immediate commencement of an enactment cannot signify an utterly novel appropriation but rather, to the contrary, already presupposes an acquisition.

What is Aristotle expressing here? More than the acquiring and losing, he is addressing having and not having a δύναμις. What does this have to do with the guiding question, which asks how a δύναμις as such is actual? To be capable of something surely means *to have* the δύναμις, and the corresponding not-having implies not being capable. This having and not-having holds the secret to the actuality and non-actuality of δύναμις. Is having thus comprehended as a kind of being? Apparently, ἔχειν and μὴ ἔχειν are understood here in a very definite sense.

Against this, the Megarians see the actuality of δύναμις in its enactment. Thus there must exist a difference between having a δύναμις and enacting it, a type of difference which of course now also reveals a coherency, since otherwise both could not be claimed as a clarification for the being present of δύνμαμις. Both determinations, ἔχειν and ἐνεργεῖν, are supposedly attributable to δύναμις. We can be more precise: According to Aristotle, δύναμις is there, is actual, if it is possessed; according to the Megarians, δύναμις is actual if it is enacted. What is in question is the actuality of δύνασθαι qua δύνασθαι.

$$\begin{array}{c} \delta\acute{\upsilon}\nu\alpha\sigma\theta\alpha\iota? \\ \diagup\diagdown \\ \delta\acute{\upsilon}\nu\alpha\mu\iota\nu\ \acute{\epsilon}\chi\epsilon\iota\nu \ —\ \acute{\epsilon}\nu\epsilon\rho\gamma\epsilon\hat{\iota}\nu \\ \text{(Aristotle)}\text{(Megarians)} \end{array}$$

b) The conflict is grounded in the Greek understanding of actuality

Aristotle and the Megarians differ in that they verify and demonstrate in different ways (one in having, the other in enactment) what they understand by the actualization and actuality of something, in this case of δύναμις. Or do they in fact understand something different in regard to the actuality of something—that is, in regard to its being present? This is the decisive point of questioning, which pervades and dominates the whole confrontation, but without becoming properly and explicitly thematized, either by Aristotle or by the Megarians. This is the basic situation, which we come across in antiquity and, from then on, in philosophy as a whole, with respect to the articulation and development of the fundamental question of philosophy: What is a being? The emerging *twofold question* in this case belongs to the fundamental question about the being and provides the inspiration for philosophizing: (1) What in general is understood by actuality (being present)? (2) What is the test and verification of this general idea of the actuality regarding a δύναμις? While the distinction between these two may not be obvious, the multiplicity of what is at stake in this question is in fact so overwhelming that at first it could deliver itself over only in one or the other of these two directions. At

§ 18. Beginning of Aristotle's confrontation

the same time, however, there is a greatness here in that this question was actually asked and not merely an afterthought.

Only through patiently laying out what is most properly at stake in the matter with regard to this theme shall we learn that the discussions in the third chapter, and thereby those of the entire treatise, are most intimately bound up with the fundamental question of philosophy. We may no longer be satisfied with extrinsically grafting the question about δύναμις and ἐνέργεια onto a schema of the different meanings of being.

Whether two different representations of actuality have emerged here or only one and the same ancient representation, namely that of *presence*, and whether this takes place with a meaning which is more narrow and focused or with a meaning which is broader, all this remains unclear. So much so that precisely the question concerning δύναμις κατὰ κίνησιν and ἐνέργεια issues in the preparation for a coming to grips with the entire question.

How do things stand with the conception of the essence of being in general, especially that of actuality, in relation to the concrete question of this chapter? Do two distinct conceptions of being present really collide, or are Aristotle and the Megarians, along with all the Greeks, here united, even to the point that only on the basis of such a unity is a conflict first able to be awakened, a conflict which only then may dispute precisely what could constitute the being present of δύναμις as such? Let us begin to anticipate what is essential here by means of statements that are necessarily general, at least initially, but which shall for their part be supported with evidence from Aristotle's concrete discussions.

Aristotle and the Megarians are completely united about what actuality in general, the being present of something, signifies; it signifies "the presence of something." This is by no means the abstract, vacuous, and flimsy explanation it might appear to be! In the process of laying out the previous chapters, we discussed at length the phenomenon of production together with the work relation for a reason, precisely in order to prepare for what now concerns us. With regard to "work," ἔργον, we made multiple differentiations: εἶδος, τέλος, πέρας, and λόγος. The work character of the work is determined by its outward appearance. And it is just this outward appearance, as something finished and intrinsically completed, which is to be brought forth, pro-duced in a single, corresponding work. To pro-duce means to make presently available (not just to make). Having been pro-duced implies first of all being finished, and secondly, and at one with this, "being at this time available." This *having been produced* is the actuality of the work; that which reveals itself in such a way "is." But precisely that which is in such a way but which at the same time does not first of all have to pass through human production, that which is essentially not in need of production, given over prior to all production—this is nature and the gods. And these are in a still more original sense; more original, this means: not essentially different. For even that which is not in need of production, and precisely this, is also understood with respect to its being in terms of the essence of having been produced. This is the sense of the basic fact that such concepts as εἶδος, τέλος, and πέρας, as fundamental moments of beings, are not restricted to things which have been produced, but rather concern the full array of beings.

Now if we say presence is having been produced, then everything which has been thus far adduced must be thought along with this in order to allow for the full significance of the fundamental Greek concept of being, οὐσία, as παρουσία, presence (and as the counter concept to ἀπουσία, absence).

This account can also immediately satisfy the dispute which lies before us, since only now do we understand the sense and validity of the Megarian thesis. A δύναμις is then actually present only if it is related to the actualization of a work. This means, however, that the δύναμις itself is enacted. For only the producing that is enacted makes evident that someone can do something and what that something is. Only here is his ability itself made available to an immediate view. In its enactment the δύναμις attests to its presence; in the enactment the δύναμις is actually present. We can and indeed must say that the Megarian interpretation of the actuality of a δύναμις is thought in a good Greek manner. Not only this, it is—up until the new advance

§ 18. Beginning of Aristotle's confrontation

made by Aristotle—the only possible interpretation of the being present of a capability.

What at first looked like an eccentric stubbornness and a deliberately strained and capricious assertion now shows itself, through a living appreciation for the Greek question of being, to be the inner necessity and the greatness of a lucid consistency (and the tenacity of fortitude) which does not let itself be led astray by mere hunches or dawning difficulties. To be sure, only in this way does Aristotle's task first receive its background and acquire its unique greatness.

Nevertheless, if one is already tacitly convinced, as is the case with our professors of philosophy and history, that here long-resolved questions are now being tossed about clumsily yet sophistically, then one will not understand why any kind of effort should be made for the interpretation of these outdated debates. But neither do we want to demand this. It would already be asking too much for these progressive men simply to confess that they do not understand what is at stake here. Indeed, these days we understand everything.

Why does Aristotle, like the Megarians, see the essence of the actuality of something in presence? This is a question which philosophy must do its utmost to answer. Here we must be satisfied with recognizing the fact and with raising the question.

We shall now attempt to determine with greater clarity the extent to which a different meaning of "presence" is nonetheless suggested for Aristotle and the Megarians.

*

At the beginning of this course, we ascertained from only general postulates that in some fashion the question τί τὸ ὄν—"what are beings?"—is necessarily being dealt with in Book Θ. We first arrived at an understanding by directing ourselves toward the fourfold πολλαχῶς of the being. The ὄν δυνάμει and ἐνεργείᾳ belonged within this as well. This, however, led back to the problem of δύναμις and ἐνέργεια. The principal thing to be explained there was what δύναμις κατὰ κίνησιν is. Now a peculiar link between δύναμις and ἐνέργεια, indeed at bottom an identification, shows itself in the Megarian thesis. The discussion of this thesis itself, the matter being treated presently, leads finally to the above-mentioned question about beings. To this question about how a capability actually is (the theme of chapter three), the Megarians answer: In its enactment. How does Aristotle meet this thesis? It seems as though he simply stretches it to impossible consequences. But in this way a clarification of the matter would not be furthered at all. Considered formally, this seems to be the case; and yet if viewed correctly, this critical confrontation does push toward the elaboration of definite central phenomena. In this way, preparations are made at the same time for a positive discussion. Thus we divide the whole into two phases. This division occurs at 1047a20. The first section articulates a series of arguments, the first of which we have examined: the ἔχειν of a δύναμις is bound to an acquisition, the μὴ ἔχειν to a giving up. Thus a mere ceasing to enact a capability is not yet no longer having it; correspondingly, beginning to enact is not identical to an utterly novel appropriation. According to Aristotle, to be capable means δύναμιν ἔχειν; according to the Megarians, in contrast, it means ἐνεργεῖν. Thus we have different answers to the question concerning the actuality of a capability. This brings us to the question: Does there exist on both sides a different conception of actuality at all, or is this conception the same and only the interpretation of the actuality of capability different? We say that the Megarians and Aristotle are united in their general conception of actuality. Both understand it as presence; οὐσία as παρουσία or, with a view to what was said earlier, having been produced. And indeed, only from this perspective does the Megarian thesis first become comprehensible; it is good Greek. And yet why is this solution, according to Aristotle's interpretation, inadequate? Why does he offer another?

§ 19. *Being in practice as the actuality of capability. The phenomena of practicing and cessation*

In the end, presence does differ according to the character of the being which supposedly is there present. Here we are dealing entirely with a being that is directly opposed to the ἔργον and its having been

§ 19. Being in practice as capability 157

produced, namely δύναμις. It is therefore manifest that the being present of δύναμις may not then be immediately taken as the presence of ἔργον, or of producing. Aristotle sees the presence of δύναμις as such in ἔχειν. What one has is, in the possession and as possessed, available, present, except that now the presence of δύναμις as such in the sense of being possessed is certainly not δύναμις in enactment.

Δύναμις as such is already actually present before the actualization, if by actualization we understand "enactment." Δύναμις ἔχειν means that something which is capable is capable in that it *"has"* a capability; it *holds itself in* this capability and *holds itself back* with this capability—and thereby precisely does not enact. This holding itself back now shows itself to us already more clearly as a way of being. This holding itself back is at the same time a *holding onto for . . .* (the enactment itself). Here we have to gather all this from the Greek word ἔχειν. —The meanings which I have designated here also come into play with the corresponding expression ἕξις.

If we are to proceed any further in thinking the confrontation between Aristotle and the Megarians, then we must now ask: How do the Megarians respond to the Aristotelian thesis concerning which the being present of δύναμις consists in being possessed? Being possessed, they might retort, is in fact something like being available. And yet let us keep our eyes open and maintain a clear sense of what being present qua presence means. Indeed, in this "merely" being possessed the δύναμις is precisely not made available, is not placed at the fore, pro-duced here, so that everyone can perceive it as actual. We have the δύναμις before us as actual only if it is enacted in our presence and in this enactment it produces and presents itself. Only in enactment does it come to light, does it "present" itself and become present. Whatever does not exhibit such a being-produced does not exist. An unenacted δύναμις is not present. Because the presence of a δύναμις means its enactment, non-enactment is equivalent to absence.

Aristotle is able to encounter this thesis only in such a way as to show that the non-enactment of a δύναμις is not already its absence and, vice versa, that enactment is not simply and solely presence. This implies fundamentally that the *essence of presence* must be understood *more fully* and *more variously* and not simply in the prevailing broad and vague generality, as with the Megarians. The essence of being requires a more original elucidation and formulation; this is the innermost tendency of the whole chapter.

We ask more exactly: Now how does Aristotle encounter the Megarian equation of the actuality of a force and the actualization of that force in enactment? By what method does he show that non-enactment is not already absence? By *pointing to a phenomenon* that first of all allows the essence of enactment to be grasped properly and thereby allows the presence of the force which lies within it to be delimited according to its own specificity. This phenomenon is that of *learning* and *unlearning* in the broadest sense. What does this have to do with our question?

Let us begin with the negative side of the Megarian thesis. According to this, the non-enactment of a capability is identical to absence, to its not being present. Assuming that this were to hold true, then the one who is capable would have to lose this capability each time it was not enacted. Non-enactment would be in itself an unlearning. And yet what does non-enactment have to do with unlearning? Just as much and as little as enactment has to do with learning. Aristotle wants to say: Let us leave this thesis to the Megarians and hold to what they draw attention to: enactment and non-enactment. And now we also want actually to hold to this and make it clear to ourselves what belongs to enactment, or, as the case may be, what does not belong to non-enactment. If, according to the Megarians, the essence of the actuality of a δύναμις lies in its enactment, then the meaning of this must lie precisely in a truly penetrating inspection of the essence of enactment and non-enactment.

Enactment is never only the emergence of something which before was completely gone, and, on the other hand, non-enactment is also not simply the disappearance of something which was there. *Enactment is practicing*, thus presence of practice and being practiced in; it is the presence of *being in practice*, the presence of something which is already present. Although enactment is presence, it is by no means the presence of what was previously simply absent but just the reverse, the presence of something which was indeed already present as well; this means, however, that this is no capricious, indefinite presence at

§ 19. Being in practice as capability

all but rather a presence which is peculiar and distinctive. On the other hand—being in practice is obviously a way of being present that does not need practicing, that is, does not necessarily have to present itself practicing. Conversely, *non-practicing* as non-enactment is not what is completely gone, simple absence; if this were so, then non-practicing would be identical to being out of practice, something which is not at all the case. On the contrary, no longer enacting, the cessation of enactment, can even signify having genuinely come into practice only now, assuming that training and being trained occurs and is cultivated only through practicing.

Thus the Megarians comprehend the essence of presence *too narrowly;* they let it be verified and presented only by that which is present in the manner of an ἔργον. They comprehend enactment merely as the emergence of something which, with and alongside the work, is present just as the work is present. Because of this narrow comprehension of presence, the essence of enactment escapes them, which, as a being-at-work, has the character of practicing. Similarly, however, this same narrow comprehension of presence closes off for them the insight that non-enactment as not practicing in itself is a way of being in practice, and therefore the presence of something; in order to be able to be one who is not practicing, I must already be proficient.

Non-enactment, ceasing to enact, is something different from the giving up and losing of a capability. Something such as this can happen through forgetting, λήθη (cf. 1047a1); the previously possessed sinks thereby into oblivion. We say: I no longer know how one does that. This no longer knowing is a no longer having expert knowledge of something; the expertise is buried. Cessation of enactment as non-enactment, however, is so little a forgetting that precisely in breaking off the "practicing" we draw the capability back into ourselves in a peculiar way; that is, we draw back into ourselves the ability to practice, the being proficient, and hold onto it especially for other situations and opportunities. Ceasing is no throwing away but rather a taking into oneself.

Ceasing, qua no longer enacting a capability, suggests a completely different way of being from what we have in mind when we observe the rain stop; here first of all, if we observe only what is present, the rain is no longer present—and that is the end of the matter. The ceasing of a capability is, on the contrary, to be taken totally differently, and at the same time is still inherently ambiguous. Here we develop this only briefly. Ceasing enactment can signify: (1) interruption—this involves more than ever a holding practice ready for later; (2) being finished—here the holding ready and conserving for something else is again what is aimed at; (3) but to abandon something, to henceforth no longer practice it—this implies: to fully withhold oneself, to withhold and withdraw the capability; this characterizes the way of being of one who leaves off doing something of which one is capable.

Non-enactment as a suspension of practice is just as little the loss of a capability. Such a loss can come about if, for example, the potter through some misfortune, πάθος, loses both hands. Then we say: For him pottery is finished. But this being finished is a totally different occurrence from, say, when the potter takes his leave from the wheel and the workplace. Indeed, even with such a loss of hands, the capability has not utterly disappeared, in the sense that the Megarians wanted to be able to assert, namely that it is simply gone. It is merely in a certain way no longer present.

Non-enactment, ceasing, therefore is ultimately also not an unlearning, because, at the very least, such unlearning requires time; "with time," χρόνῳ, we get out of practice. But ceasing to practice in the sense of breaking off never happens through time, but rather always at a definite moment, to which it could never be attributed that through this we are already out of practice.

The being present of a capability is to be understood as being in practice; as such, it expresses precisely the ownmost actuality of capability as capability. This being is certainly the non-enactment of a capability, and yet non-enactment is not the loss of capability. For this a totally different character of being and occurrence is required (λήθη, πάθος, χρόνος). However, to the extent that the presence of a capability as such hardly depends upon enactment or non-enactment, just as little does it depend upon the being present or not-being present of the work to be produced in the enactment, or even of the work already produced in the enactment or, more exactly, of that with

§ 19. Being in practice as capability

which the δύναμις occupies itself and to which it inherently refers—the πρᾶγμα (1047a2). The actuality of the δύναμις as such remains completely independent of the actuality of that of which it is capable—whether it has actually been produced, or is only half finished, or even not yet begun. We intentionally say: independent of the actuality of that of which the capability is capable. This is so because it belongs to a capability as actually present that it be capable of something, as the realm and reach of this capability—although admittedly in such a way that it now remains undecided what actualizes itself in this, and how and to what extent it actualizes itself.

Even if that with which the capability occupies itself (πρᾶγμα) is not yet produced, the capability as such is already actual. And likewise, the capability remains actually present and does not disappear, even if the matter with which it occupies itself happens to disappear, assuming, that is, that this matter is already present or even finished (which is what the εἰ γάρ ἐστιν signifies). Thus the thought at 1047a2 is completely clear; the ἀεί makes no sense at all, and even less do the forced explanations that interpreters have grafted onto it.

If we do pay attention not only to whether and to what extent Aristotle, in the passage that has now been laid out, refutes the Megarians, but first of all put into relief what Aristotle thereby brings up in a positive way, then several things come to light: (1) Aristotle wants to bring into view for the very first time the proper manner of being actual of a δύναμις; this occurs through the emphasis upon δύναμιν ἔχειν, having a capability. (2) It must be shown that this having has gone through a training (cf. ἔχει, 1047a3); "having" is a presiding over the capability and, in this sense, a being trained. (3) It must be seen, however, that this training is to be differentiated from the practicing—from the enactment—which first becomes possible because of the training. (4) The peculiarity of enactment as practicing first becomes comprehensible on the basis of the phenomenon of practice, of being in practice, and thus on the basis of the correct insight into the kind of actuality of capability as capability.

Aristotle does not deny ἐνεργεῖν, being at work, as one way in which δύναμις is actual. But he does deny that this is the single, basic way in which the actuality of a force is. On the contrary, ἐνέργεια—

ἐνέργεια κατὰ κίνησιν—is first comprehensible in contradistinction to being practiced. From the very outset and for the purpose of the whole chapter, it is to be maintained that Aristotle, in resisting the equation of the actualization of a force in enactment with the actuality of a force, does not dismiss ἐνεργεῖν but rather wants simply to restrict it to its own domain, to delimit it, and thereby to define it.

This means, however, that precisely ἐνέργεια κατὰ κίνησιν is also a central problem, if not even the central problem in chapter three. And yet according to the questions which are posed, the topic is not at all ἐνέργεια ἐπὶ πλέον. Indeed, the transition to this is to be prepared here. And we do not want to forestall this. Instead, we want to sharpen our view of the inner train of thought of this great chapter. For this reason, we intentionally lingered a little longer with Aristotle's first argument, but all the better since here the arguments which follow move in the same general direction.

At the risk that you have already grown weary of what is being discussed (presumably because you have already long since understood it), I shall repeat the presentation of what is decisively at issue in order that you may learn to become increasingly more tactful with the obvious and to sense its innermost questionability. The actuality of a capability does not consist in its actualization. The discussion of the actualization of a capability is thoroughly ambiguous. "A capability for something actualizes itself" can mean: (1) That which beforehand was not there, came to be; for example, one becomes trained in pottery, something which one previously was not. The capability is in this sense actual, as well, and even more so if it is not actualized in the second sense—which means (2) if the training is employed and enacted. (3) A capability is actualized if that of which it is capable becomes itself finished and is produced; for example, the mug as something available and present. Between these three fundamental meanings of the "actualization of a capability," peculiar relationships result, to which I have yet to return.

✦

The thesis of the Megarians is not at all as arbitrary and strange as it at first appears—especially if viewed in light of the ancient under-

§ *19. Being in practice as capability* 163

standing of being: The actuality of a capability lies in its enactment, in which it presents and produces itself. To which Aristotle replies: The actuality of a capability as capability lies in its ἔχειν, in having. The intent of the Aristotelian refutation is not just to show that the actuality of a capability does not lie in its enactment, but rather he basically wants at the same time to demonstrate that the Megarians posit actuality in this only because they are not capable of adequately comprehending the essence of enactment itself. This involves introducing into the inquiry a heretofore overlooked phenomenon. Enactment is practicing, non-enactment is non-practicing—but not being out of practice, the loss of a capability. To such losing belongs something totally different from what belongs to the suspension of practicing (cf. the remarks concerning λήθη, πάθος, and χρόνος). Ceasing is thus not a giving up, but rather a taking into oneself, in accordance with ἔχειν as holding oneself back, withholding, and holding onto for . . . The actuality of a δύναμις is, however, independent not only of the enactment, but rather also of the presence of the πρᾶγμα (or of the ἔργον). "Actualization of a capability" itself has a multiple meaning: (1) the capability is simply there; (2) it is in its enactment; (3) it presents itself by means of that which it produces in the enactment.

Peculiar determining relationships exist between these three kinds of actualization. A capability can be actual in the second sense, that is, in enactment, without its being actual in the third sense, thus without having actualized itself or, as we also say, without its having left its mark in a work. The actualization of a capability in the third sense, however, presupposes the actuality in the first sense as well as necessarily having gone through the second sense. In contrast, the actuality of a capability in the first sense, being trained, is not dependent upon the second and third sense.

But this requires more scrutiny. Someone could be actually skilled as a potter without exercising the τέχνη, and without mugs produced by him being at hand. The being actual of the capability does not consist *in* the actualization in the second and third sense. But does it not nevertheless consist "*of*" these? How can one be *trained* without having *practiced?* Training develops through practice. And practicing is actual and itself, when it follows through on what belongs to it to the end, when it has actually brought about a work and this work "stands." This is just an indication that, although being trained does not consist in practicing, still, the connection between the two has a peculiar character.

But let us return once again to the Megarian thesis and consider its motive and foundation, which in a certain way are legitimate, for otherwise it would not be worth mentioning nor would it be in need of revision. What is peculiar about the Megarian thesis, and this must now be made sufficiently clear, does not just consist in the fact that the Megarians posit the actuality of a capability exclusively in its enactment, but rather in that they comprehend this enactment itself one-sidedly and not in its full essence.

Enactment is indeed presence and non-enactment absence, but these statements do not hold simply in a straightforward way. Enactment is rather a practicing and as such, if it is at all, the presence of training. But this explanation too remains a poor one, since we can easily see and respond right away that the presence of being trained does not especially need enactment in order to be. And why not? What is lacking? The being trained which is present [*anwesend*] is first a practicing only if it has come into such practicing. But being trained, however, "comes" to such practicing only if it passes over into it and is transferred to it. With this, being trained is not transported to something else, which in each case would be at hand, but rather the being trained passes over beyond itself into something which first forms itself only in and through the passing, what we call carrying out and practicing. This *passing over beyond itself* of a capability into enactment requires its own room for free play. This free play is given, however, through the reach of capability that was characterized earlier (cf. p. 96ff. and p. 112ff.). For through this reach every capability as such masters something, a mastery which pre-forms itself along with that which can be mastered, and in such a way that while practicing it informs itself in the practice and thereby alters itself.

This is an entirely essential moment of the actualization of a capability, set apart from every other kind of actualization. If some possible thing, for example, a table, becomes actual, this means that this thing represented generally is actualized and becomes present precisely as the here and now. But if, in contrast, a capability is actualized, then this capability itself does not become actual like the possible thing, but instead what then becomes actualized, as that which the capability actualized through actualizing itself, is the other of this capability itself.

Carrying out and practicing can then be only in that the capability begins by way of passing over beyond itself. A non-enacted capability is therefore actual in such a way that a *not-yet-beginning* belongs to its actuality positively, something which we previously approached as holding oneself back. The passing over and beyond is not just forced onto a capability as something new, but rather it is something held onto in this holding itself back.

At the very outset, therefore, Aristotle is aiming at this phenomenon of passing beyond and going over when in the discussion he points to beginning and ceasing and to learning and unlearning. This does not here concern the platitude that with any enactment at all it is necessary for it to have begun at some time, but rather it is the "how" of the beginning which is essential here, and the origin of this "how" precisely from out of capability.

We have constantly to recall these phenomena to ourselves anew and to let ourselves be exposed to the complete wonder which they hold. If we thereby go beyond Aristotle, this is not done in order to improve upon what is said, but primarily to begin simply to understand. With this it matters little which manner and form of expression Aristotle for his part happens to use in carrying out these necessary considerations. An understanding of the following passage depends entirely upon the degree of thoughtful perseverance regarding the overall connections between the phenomena under discussion.

§ 20. *The actuality of the perceptible and the actuality of the capability of perception*

a) The problem of the perceptible and the principle of Protagoras

1047a4-7: καὶ τὰ ἄψυχα δὴ ὁμοίως· οὔτε γὰρ ψυχρὸν οὔτε θερμὸν οὔτε γλυκὺ οὔτε ὅλως αἰσθητὸν οὐδὲν ἔσται μὴ αἰσθανομένων· ὥστε τὸν Πρωταγόρου λόγον συμβήσεται λέγειν αὐτοῖς.

"And with soulless beings it is therefore likewise arranged; something will be neither cold nor warm, nor sweet, nor perceptible at all, if perceiving is not in practice; thus it necessarily turns out that they (the Megarians) concur with the doctrine of Protagoras."

This sentence supposedly presents another argument against the Megarian thesis. It is concisely formulated, almost like an aside, more a reference to what is well known and repeatedly discussed than an explanation. This does not allow us to conclude that an argument of lesser importance is being brought forth here, but quite the contrary: here we have a signaling toward a nexus of questions, which is not only highly significant in the later portions of this whole treatise, but which in general assumes a prominent role in the debate over the fundamental questions of ancient philosophy.

The form of argumentation is the same, namely, once again a reference to συμβαίνειν—the certification of an inherently impossible consequence which results from assuming the Megarian thesis. We encounter the unacceptable result this time as the teaching of Protagoras. The mention of this philosopher or Sophist and his teaching provides an important clue to the general area of questioning in which we have to pose the presented argument. Stated more precisely, we gain a handle on how we have to understand what Aristotle here calls τὰ ἄψυχα.

To be sure, this appears at first to be quite clear. In the previous argument, the topic of discussion concerned the τέχναι; these are, as we know above all from chapter two, the δυνάμεις μετὰ λόγου (or else the ἔμψυχα). Now comes "a" deliberation, and just as the ὁμοίως suggests, the same deliberation but in relation to ἄψυχα. We saw indeed that Aristotle divides the possible δυνάμεις into two regions, the ἔμψυχον and ἄψυχον. It must now be asked: How do things stand with the actuality of forces in the region of soulless beings? Does the actuality of these beings consist in acting, as the Megarians insist, or are they properly actual precisely in their power to do something when not acting—but poised on the edge, perhaps like an accumulated force (so-called potential energy)? In line with this question, one must understand the καὶ τὰ ἄψυχα δὴ ὁμοίως. What comes next in unfolding the corresponding question in relation to soulless beings is then evi-

§ 20. Actuality of the perceptible

dent: it is to be shown how, say, a merely material thing can affect another thing of the same kind—how, perhaps, to pick up an earlier example, a warmer body warms another. The topic is in fact ψυχρόν and θερμόν. So the question to be discussed would be whether the being warm of the warming body consists only in its actually giving off warmth to another body, or whether the warm body also exists precisely as a body capable of warming when it is not warming another body.

But this is not at all what is under discussion in the text. What is in fact being dealt with here is the guiding question in general, namely the question concerning the actual being of what is capable as such, the εἶναι of δύνασθαι, or the question of how the εἶναι in δυνατὸν εἶναι is to be understood. Only now it is undecided which δύναμις is asked about in this way. The previous argument dealt with such a δύναμις μετὰ λόγου. And so is the concern now, where ἄψυχα are indicated, δύναμις ἄνευ λόγου? But we see that the discussion is not about this; hence it is neither about τέχνη nor about δύναμις ἄλογος. Are there, then, still other δυνάμεις? Let us look at the text.

Aristotle says that on the assumption of the Megarian thesis, the perceptible is actual only during the enactment of perceiving. Thus this deals with the enactment and actuality of αἴσθησις. And αἴσθησις is in fact a δύναμις and is indeed expressly formulated as such. It is a δύναμις which belongs to ψυχή, namely to ψυχή τῶν ζῴων (cf. *De an.*, Γ 9, beginning); νοῦς and διάνοια are correspondingly θεωρητικὴ δύναμις (Β 2, 413b25), as the capability for simple observing. Thus we have a δύναμις, and one even explicitly given as an ἔμψυχον; but at the same time it is no τέχνη, no capability for producing an independently present work. Then what is it? Αἴσθησις is a capability for ἀληθεύειν, for making manifest and holding open, a capability for knowledge in the broadest sense. And yet—if this deals with an indubitable capability of the soul, what does it then mean that the whole argument is introduced with καὶ τὰ ἄψυχα ὁμοίως, where it is unambiguously stated that soulless beings are now being dealt with?

What is being dealt with is in fact both ἄψυχον and ἔμψυχον, yet not as two adjacent realms which have their corresponding δυνάμεις within them; rather, what is being dealt with here is that very δύναμις which, according to its essence, is nothing other than *the connection and relationship of one definitively constituted* ἔμψυχον *with a determinate* ἄψυχον *that is apprehended in a specific way*. Let us note well: I am not simply saying that what is being dealt with here is a δύναμις that in itself makes up the relationship of the ἔμψυχον to the ἄψυχον, since a plant is also an ἔμψυχον but does not have the δύναμις being dealt with here, namely αἴσθησις. Admittedly Aristotle says of this in one and the same passage (B 12, 424a27f.; cf. above, p. 108) that this is λόγος τις καὶ δύναμις—as δύναμις something like a relationship to . . . another. But again neither is it a relationship to just any ἄψυχον in just any way but to the ἄψυχον with the character introduced in the text (*Met.* Θ 3, 1047a5): θερμόν, ψυχρόν . . . —warm, cold, sweet, colorful, sonorous, fragrant; such ἄψυχα therefore are accordingly taken in a determinate way, as αἰσθητά, as the perceptible.

This kind of δύναμις, namely αἴσθησις, is indeed, as is the case with every δύναμις, in relation to something as a power to do something. However, this relation here in the case of αἴσθησις is one which is entirely distinctive and unique. This implies now that what is referable to such δυνάμεις also has its own unique character. Thus, for example, it may not be equated with that to which a τέχνη relates, the ἔργον. More precisely: the ἔργον of αἴσθησις as a δύναμις is not a thing which has been produced and is not at hand *as* something produced and finished. We do not produce things through perceiving; we do not, for example, produce something like a colored thing—this we accomplish by painting; nor do we produce this or that tonality—this we accomplish through the tightening and strumming of strings. The ἔργον of αἴσθησις, just like that of νόησις, is ἀλήθεια—the openness of beings, and in a special manner the perceptibility of things—namely those that show themselves to us in their coloredness, in their tonality.

We are discussing αἴσθησις as δύναμις. Αἴσθησις is a relationship of that which opens to that which can take part in such openness, that is, to beings in their particular manner of being, or to their being in general. Αἴσθησις as a relationship so characterized also defines the

§ 20. Actuality of the perceptible

living beings that we call human. Perception is also a capability of the human. It would be erroneous to hold that the human then possesses in addition to this the property of thinking and of reason, such that we have only to take this away in order to have what the animal has. The perceiving of the animal is rather from the ground up other than that of the human. Humans comport themselves perceptually *toward beings*, something of which the animal is never capable, even when the animal can perceive incomparably more keenly than humans, as is the case, for example, with the eagle in regard to sight. In this perceptual relation, the relationship of the human to beings and of beings to the human is in a certain way co-determined.

With the new argument, which at 1047a4 is so oddly and apparently so misleadingly introduced with καὶ τὰ ἄψυχα, we are in fact thereby ushered into that very nexus which represents the central point of the teaching of Protagoras. Aristotle quite often finds the occasion to speak of this teaching, most thoroughly and pointedly in the refutation that is found in *Met*. Γ 5. But Plato, too, more than once clarified and secured his own views in and through a confrontation with Protagoras. Thus the first main section of the dialogue *Theatetus*, which has as its theme ἐπιστήμη τί ἐστιν—What is knowledge?—is entirely dedicated to the confrontation with Protagoras. This shall serve as an essential source for us. Here as well we find the general principle of Protagoras introduced (*Theatetus* 152a): φησὶ γάρ που "πάντων χρημάτων μέτρον" ἄνθρωπον εἶναι, "τῶν μὲν ὄντων ὡς ἔστι, τῶν δὲ μὴ ὄντων ὡς οὐχ ἔστιν." "The human being is the measure of all things, of beings, that they are, of non-beings, that they are not." And this principle is based on the essence of αἴσθησις, 152a6ff.: οἷα μὲν ἕκαστα ἐμοὶ φαίνεται τοιαῦτα μὲν ἔστιν ἐμοί, οἷα δὲ σοί, τοιαῦτα δὲ αὖ σοί. "However each thing shows itself to me, so it is for me, but howsoever to you, so it is in turn for you." And further on (b11): τὸ δέ γε "φαίνεται" αἰσθάνεσθαί ἐστιν. "The 'it shows itself' means nothing other than: it is perceived, manifest in perception." (That is, in connection with the general Protagorean conception of ἐπιστήμη.) So much for a most general orientation concerning the teaching of Protagoras.

At the same time, however, it must be emphatically stressed that it is not easy, particularly in the Platonic discussions, to distinguish the most genuine opinion of Protagoras from what Plato interpolates by means of inferences and complications. It is thus not immediately clear what Protagoras meant with his ἀλήθεια; as a precaution one must be on guard against interpreting this Protagorean teaching in a crude, so-called sensualistic sense and labeling it an epistemological school of thought, about which it could be convincingly demonstrated with a turn of the hand to any halfwit that such a doctrine leads to so-called skepticism. For if the true is at any time precisely that which appears to someone in the manner that it appears, and is true only for this reason, then of course a generally valid and objective truth would not be possible. We do not here want to discuss any further such supremely reasonable argumentation; only one thing must still be pointed out: Such argumentation is built upon the assumption that truth would not be truth unless it holds for everyone. But this assumption is without grounds, or else it is not at all made clear what it would mean to ground this assumption. One forgets to ask whether the genuine essence of truth does not consist in the fact that it is not valid for everyone—and that the truths of everyone are the most trifling of what can be gleaned from the domain of truth. But if one ponders and questions in this way, then it becomes possible for Protagoras's oft-maligned statement, which can be misappropriated by anyone new to philosophy, to hold a great truth, and indeed ultimately one of the most fundamental truths. Admittedly not just anyone can see this unconditionally, but rather only the individual as an individual is capable of gaining this insight each for him- or herself, assuming that this individual philosophizes.

It is no accident that already in antiquity the principle of Protagoras was indeed given a very specific interpretation by Plato and Aristotle which allowed skeptical conclusions to be drawn from it. Here it may be relevant that the one predominant aspect handed down to us formulates the essence of knowledge in terms of perception. But the Aristotelian confrontation in *Met.* Γ 5 already clearly betrays that something more essential lies behind this teaching which, owing to the overwhelming significance of Plato and Aristotle, is all too easily neglected in the popular assessment.

To be sure, only this one question can occupy us now: What does the question concerning the kind of actuality of δύναμις qua δύναμις

§ 20. Actuality of the perceptible

have to do with the principle of Protagoras? More exactly: To what extent does the Megarian thesis in regard to the δύναμις that has the character of αἴσθησις, a thesis which searches for the being at hand of a δύναμις in ἐνεργεῖν, lead to the teaching of Protagoras, conceived as the denial of the possible knowledge of beings themselves?

*

According to the teaching of Protagoras, it always is only what is just perceived which is, and in each case it is only such as it is perceived. Thus we never know beings as they are in themselves, as unperceived, as not formed in a perception. Different humans can never reach an agreement over one and the same being, since each views it in his or her own fashion. What is warm to one is cold to another. Indeed, the very same human first finds something sour, then tart, then sweet, according to his or her own bodily condition. If such conclusions have already been drawn from the teaching of Protagoras, then let us push this still further. Not only can different humans not reach agreement over the same issue, strictly speaking, they cannot even once be in disagreement over this same issue. Thus neither is there a possibility for conflict, since this evidently presupposes something selfsame and perceived, which is considered by many as simply one and the same, the same something which one perhaps speaks for, but another against. And again, the broader underlying presupposition that is operative here is that there be a *perceptible being* at all.

What does this call for? Nothing less than such a being which itself and from out of itself, prior to all being perceived, is empowered (δυνατόν) to be perceived. This perceptible being—that is, a being with the ability to be perceived—must "be" as this being with this ability, that is, it must "be" actual, if a perceiving and becoming manifest is to occur at all. (See Kant's solution for the possibility of this "being"—the event of objectivity.)

If the Megarian thesis holds, then the actuality of such a being, the perceptible as such, is undermined. How so? If the actuality of that which is empowered and capable of something lies in its enactment, then the perceptible as such "is" actual if and only if and precisely only so long as it is perceived or, as we say, only if it is "actually" being perceived. Thus, Aristotle concludes further, without the enactment of perception, such things as a colored thing and color, or a sonorous thing and tone would not be present. The variegated multiplicity of this immediate, sensually given world must start anew with each and every enactment of perception and then once again desist, so that what emerges and passes away there is itself nothing inherently present. Moreover, not only must the Megarians as a consequence of their thesis arrive at this conception of what is immediately and actually given, they must in general deny the possibility of a being that is in and of itself present, since this can be granted only with the acknowledgment that the being present of something that is perceptible does not remain singularly dependent upon the enactment of perception. In this way it turns out that the Megarian thesis in its implications goes much further still and grasps the essential conditions for the possibility of what is perceptible. It does not just appeal to the factically existent difference of each differently perceiving human. All the more so, then, and all the more certainly do the Megarians have to arrive at the teaching of Protagoras.

Aristotle thus pushes the Megarians and the implications of their thesis to a point which, as the invocation of Protagoras suggests, is just as much the dissolution of the possibility for truth as it is the dissolution of the self-reliant actuality of what is present; the latter is taken thereby to suggest that the currently discussed argument concerns the ἄψυχα. The actuality of what is present as the actuality of something self-reliant then still remains intelligible only if it can be shown *that the actuality of what is perceptible as such does not lie in enactment of perception.*

With this a task is posed which Aristotle does not positively resolve but rather exhibits in its inexorability. The entire subsequent history of philosophy, however, testifies to how little the solving of this task has met with success. The reason for this failure has little to do with not finding a way to an answer, but much more with the fact that continually and up until the present day the question as such has been taken too lightly. Here we will have to dispense both with unfolding this question in its many-sidedness and with showing thereby how

§ 20. Actuality of the perceptible 173

something essential is lacking in Aristotle and in antiquity in general. But we shall forgo this in order to bring the question as a question into its own. And yet, on the other hand, in connection with our guiding theme it was precisely through Aristotle that a decisive step was taken toward the proper formulation of this question.

That which has obstructed the proper formulation of the question has confronted us throughout this entire consideration, without our actually grasping it clearly enough. It is nothing other than the double character of our theme, which has made itself apparent in the presently discussed argument. The discussion is to be about the ἄψυχα, then it is not, but instead about αἴσθησις as δύναμις. But then again, neither is this what is under discussion, if this is simply an ἔμψυχον in the sense of something present in the soul. The topic is rather the ἄψυχα qua αἰσθητά—and it is αἴσθησις qua αἰσθάνεσθαι τὰ ἄψυχα. What is in question is not how soulless material things at hand exist among themselves in relation to each other, but rather how they can be manifest in themselves as beings in themselves without being infringed upon by the fact that the occurrence of this being manifest is bound in itself to the actuality of the besouled, that is, to the actuality of human beings.

Aristotle was not capable of comprehending, no less than anyone before or after him, the proper essence and being of that which makes up this *between*—between αἰσθητόν as such and αἴσθησις as such—and which in itself brings about the very wonder that, although it is related to self-reliant beings, it does not through this relation take their self-reliance away, but rather precisely makes it possible for such being to secure this self-reliance in the truth.

But this requires that it simply be possible for us to understand something as actually present, even and especially when this present being is present as something able to be this or that, in this case as what is able to be perceived. (This is the possible belonging to the world of beings, in which they first "become" beings and thus make themselves apparent as something which before this appearance also was not nothing.) The independence of things at hand from humans is not altered through the fact that this very independence as such is possible only if humans exist. The being in themselves of things not only becomes unexplainable without the existence of humans, it becomes utterly meaningless; but this does not mean that the things themselves are dependent upon humans.

In order, however, actually to bring out these fundamental relationships and truths with complete clarity, and this means to delineate them thoroughly, and above all to circumscribe their boundaries and type of certainty, it would require again the entire effort of a philosophy. Only in this way can we ever fathom the whole site on which we as humans stand. And only when we have traversed this site in its entire situatedness are we capable of deciding with clarity what ἄτοπος is—to be without site and not able to be accommodated within that which can have a site at all.

Understood in this way, the statement by Protagoras takes on an entirely new meaning, namely, one which raises it to the most lofty principle of all philosophizing. "The human is the measure of all things, of beings that they are, of non-beings that they are not." A fundamental principle—not a cheap and easily accessible assertion, but an initiation and a staking out of the question in which the human goes to the very basis of its own essence. This questioning is, however, the basic activity of philosophizing.

The now-clarified argument of Aristotle which related to the ἄψυχα points to a fundamental question of philosophy. From this we must infer in which essential nexus this thematic question in general moves.

b) The practicing and not-practicing of perception

How much we must comprehend the ἄψυχα as αἰσθητά, in their being perceived, but to be sure not in such a way that αἴσθησις then comes into question in an isolated and detached manner, is demonstrated in the following argument (*Met*. Θ 3):

> 1047a7-10: ἀλλὰ μὴν οὐδ᾽ αἴσθησιν ἕξει οὐδὲν ἐὰν μὴ αἰσθάνηται μηδ᾽ ἐνεργῇ. εἰ οὖν τυφλὸν τὸ μὴ ἔχον ὄψιν, πεφυκὸς δὲ καὶ ὅτε πέφυκε [καὶ] ἔτι ὥς [I am reading ὥς instead of ὄν, H.], οἱ αὐτοὶ τυφλοὶ ἔσονται πολλάκις τῆς ἡμέρας καὶ κωφοί.

"Indeed (a living being) could not even 'have' perception if it were

§ 20. *Actuality of the perceptible* 175

not (as long as it was not) engaged in perceiving, if it were not at work. Now if what is blind is that which does not have sight (the capability for seeing) when this is appropriate to its nature, at that time and further in that manner that it is so appropriate, then would the same humans repeatedly need to go blind during the day (and likewise deaf [probably an addition, (H.)])."

This argument is introduced by means of an intensified comparison: ἀλλὰ μὴν οὐδέ. Here we find the relation to what came earlier. There it says that if the actuality of that which is capable as such lies in its enactment, and thus if the perceptibility of what is perceptible lies in its being perceived, then there would be no perceptible being, nothing of the sort that we could also simply represent as self-reliant in itself. There would be, nonetheless, according to the intermediate thought—perceiving. But even this, continues Aristotle, could not be actual unless it were constantly engaged in its enactment.

Now we see that in fact αἴσθησις is especially being asked about, that is, *perceiving* and its actuality as δύναμις, but nevertheless without its very essential relation to the αἰσθητόν being kept in view. But this of course expresses the proper character of this δύναμις insofar as it is not a τέχνη, not an ἔργον produced as a being at hand for itself; and yet here too the topic concerns ἐνεργεῖν, being at work in the sense of the enactment of that for which a δύναμις is capable. Here being at work is ἀληθεύειν: taking from concealment as taking-for-true, perceiving [*wahr-nehmen*]. Ἐνεργεῖν and ἐνέργεια no longer have here the originally very narrow reference to ἔργον, but nevertheless they still have the meaning of enactment.

With respect to this δύναμις, αἴσθησις, the ἔργον of which is ἀλήθεια, openness, we now proceed along the corresponding path to a conclusion which demonstrates its own impossibility, in that it comes up against strong and incontrovertible facts of our Dasein. In the case of τέχνη the builder had to cease being a builder when he stopped building. Correspondingly, a perceiving with the eyes, for example, would now have to be no longer, just as soon as it is not expressly enacted, as soon as, say, the eyes are closed (then we are no longer perceiving). Since according to the Megarians there is no proper actuality of capability as such, non-perceiving can signify only no longer being able to see, in the sense that the capability is utterly gone and is not.

Considered in just this way, Aristotle may not even argue with recourse to those who have gone blind, as he does, since the theory of the Megarians does not even admit that someone might be blind. For *to be* actually *blind* means in fact actually being one who is unable to see—thus a determinate, privative mode of being able. Being one who is "actually unable," actually being such a being, is fundamentally distinct from being not at all able, as is the case, say, with a piece of wood. A piece of wood would need precisely to be able to have the δύναμις for sight, would need to be characterized by an aptitude and a power to do something, in order just to be able to remain blind. And Aristotle does sufficiently indicate that a difference clearly exists between the suspension of perceiving and going blind.

Between actually seeing and being blind lies not-seeing in the sense of the non-enactment of visual perception, a non-enactment which inherently and actually is "able to enact at any time." Because the Megarians cannot reconcile themselves to this fact, they are compelled to portray the transition from actual non-perceiving to perceiving as a transitionless exchange between being blind and being able to see. Strictly speaking, as I said, the jump necessarily given here at all times implies still another: that of the constant interchange from the stone or wood or something similar to the animal or human, or the reverse, the return from the latter to the former.

And yet what Aristotle does not discuss is the *connection* between this third argument and the earlier specified second argument. Indeed, this also goes beyond the immediate purpose of these arguments. On the other hand, it is precisely from out of this connection that the possibility for a positive clarification of the question that we touched upon becomes visible. What I mean by this is that conceiving of θερμόν and so on as αἰσθητά is grounded in the very fact that perceiving also can still be there in the manner of a not-yet- or no-longer-enacting ability. The non-enactment of something such as perceiving does not imply its utter lack. The ability to enact as something that is there, however, is the very disposition within which what could be perceived or what has been perceived is represented, but precisely by

§ 21. *Conclusion of the confrontation* 177

a kind of knowing in which what is in this way perceptible is not entirely relegated to a constantly being perceived. Drawing oneself back out of the practice of perceiving is not the mere breaking off and disappearance of this practice, but rather has the character of a giving over of the perceived to itself as something which is then perceivable. Thus it can and must be asserted that *the self-sufficient actuality of what is perceptible is not at all experienced fundamentally in the occasional actual enactment of perception but instead first in its unique no-longer-enactment and not-yet-enactment*. (This points to an inner connection between truth and time.) Only in terms of the phenomenon which was indirectly necessitated by the last argument concerning δύναμις, namely, the ability to perceive as something unactualized but nevertheless actual, does it first become possible for the αἰσθητόν to be released as something which from out of itself can offer itself to being perceived; within this suitability for being affected through openness there lies the indication of its genuine independence.

Thus far Aristotle's argumentation against the Megarians concerns three points: (1) the δύναμις μετὰ λόγου (τέχνη); (2) the δυνατόν of a δύναμις in the sense of αἴσθησις (the αἰσθητόν); (3) αἴσθησις itself. For each a twofold indication was made: (a) the thesis of the Megarians leads to impossible conclusions with regard to the phenomenon which each time was discussed; (b) the phenomena themselves are not comprehended in their genuine essence; they are not granted their full content. Therefore, the Megarian thesis is not only untenable but, in terms of the whole matter, insufficient as well.

§ 21. *The conclusion of the confrontation: the Megarians miss the movement of transition which belongs to a capability*

As the ἔτι at 1047a10 now betrays, the Aristotelian argumentation has not yet come to a close. It still remains to be asked: Does this ἔτι, "further," provide a juncture for only a continuation of the argument in the same direction, and so in relation to the phenomena already dealt with and oriented toward the general thesis of the Megarians which we cited? Or is the argumentation in terms of its content a different one, precisely because now a new argument of the Megarians is introduced? The latter is the case.

1047a10-13: ἔτι εἰ ἀδύνατον τὸ ἐστερημένον δυνάμεως, τὸ μὴ γενόμενον ἀδύνατον ἔσται γενέσθαι, τὸ δ'ἀδύνατον γενέσθαι ὁ λέγων ἢ εἶναι ἢ ἔσεσθαι ψεύσεται· τὸ γὰρ ἀδύνατον τοῦτο ἐσήμαινεν.

"Further, if an incapable being is one from which capability has withdrawn, then that which has not reached enactment (μὴ γενόμενον) must be incapable of coming to enactment; and yet whoever claims that something incapable of coming to enactment is or will be, that one lies to himself, since to be incapable means just that."

Once again, according to the manner of proceeding, the argument works up to an impossible conclusion; this time it is to lie to oneself: to give and assert as true that which at the same time is known to be untrue. According to the content, however, there is a difference over and against the preceding arguments. The difference is twofold: (1) Now it is not δύναμις, δυνατόν, and δύνασθαι which have entered the discussion, but rather ἀδύνατον εἶναι; (2) this, however, in such a manner that simultaneously another argument of the Megarians is introduced. The position of εἰ ἀδύνατον ... ἔσται is to be understood in this way. Not only does this result from the whole form of the sentence, but it can be directly demonstrated that Aristotle here brings an explicit and completely central argument of the Megarians into the discussion.

*

In this passage at 1047a10ff., we find the content of a Megarian principle, which by virtue of its meaning had repercussions during the later period of the school. It provided the context for one of the famous proofs in the Hellenistic era by the Megarian Diodoros. On account of the irrefutability of this proof it "remained master" and therefore carried the name κυριεύων.[1] The content of this λόγος

1. See Ἀρριάνου τῶν Ἐπικτήτου διατριβῶν βιβλία τέσσαρα, Book II, Chap. 19, beginning. (*Epicteti Dissertationes ab Arriano digestai* . . . , iterum rec. H. Schenkl; Leipzig, 1916).

§ 21. Conclusion of the confrontation

κυριεύων is briefly the following: "If something would be possible which neither is nor will be, then something impossible would proceed from something possible; but something impossible cannot proceed from something possible. Thus nothing is possible which neither is nor will be."[2] The opening clause in this proof (if something would be possible which neither is nor will be, then something impossible would proceed from something possible) found support in the thesis: πᾶν παρεληλυθὸς ἀληθὲς ἀναγκαῖον εἶναι (loc. cit. in Epictetus). If with two contrary cases the one occurs, if the one thing which is possible (δυνατόν) becomes actual, then the other possible thing becomes impossible (ἀδύνατον), since "everything that has happened once is necessary," that is, has necessarily come to pass. That which has not come to enactment is now impossible and cannot therefore have been possible earlier, for if it were otherwise, an impossible something would then have arisen from something possible, according to Diodoros.

As the translation of δυνατόν and ἀδύνατον as "possible" and "impossible" already indicates, we find here—and this can be demonstrated by the content of the whole argumentation—that the meaning of capability, having-power-to, and the corresponding privative meaning is confused with the meaning of possible and impossible. Something like this apparently must also have been set forth in the Megarian debate of the earlier period. Of course, we have to note that at that time both meanings could not yet have been mixed together, because they had not yet even been separated. The clear division between the two and the simultaneous "derivation" of one from the other was accomplished for the very first time in this Aristotelian treatise.

We wove the reference to Diodoros and his argument into our discussion in order to demonstrate that here it is still readily apparent what is being clearly expressed in this Aristotelian text (1047a11) as the decisive feature of the Megarian argumentation: the μὴ γενόμενον —that which has not come to enactment. Everything which has not

---

2. See E. Zeller, *Über den* κυριεύων *des Megarikers Diodorus.* Proceedings of the Royal Prussian Academy of Science in Berlin (Berlin, 1882), pp. 151-59, p.153.

come to enactment is the non-actual. But since, according to the general thesis, something capable is only as something actualized, that which has not come to enactment is at the same time something incapable. The Megarians do not allow the non-enacted to remain as something capable, as a being with capability, but instead, according to their thesis, they must address the non-enacted as a being without capability. But here Aristotle responds that this does not amount to asserting that something which has not become and has not come to enactment is incapable or would be incapable, since to be incapable in itself means just this: not being at all. Thus in order for the Megarians to demonstrate on their own terms that something which has not come into being would be an incapable being—and this means for them what is not at all actual—they require not only the principle that all things past are necessary, but rather also that the incapable is simply and all the more so the non-actual. This is so because the capable is what it is and can be what it is only as the actual.

Thus we too can easily see here where the relevant inadequacy of the Megarian conception of the being present of a capability lies—in that they see in the phenomenon of incapability only the mere negation of being present, the negation of enactment as presence. They have no vision for the fact that the incapable is actual precisely because it *does not find the transition* to enactment. To not find the transition to . . . : this is not nothing, but instead can have the pressing force and actuality of the greatest plight and so be what is properly urgent.

From this the twofold mistake in the phenomenon becomes still clearer: (1) The Megarians comprehend the "non" as pure negation—rather than as a distinctive privation. (2) That which is negated, enactment itself, they comprehend only as the presence of something—rather than as transition, that is, as κίνησις.

Now Aristotle brings all of the discussed arguments against the Megarians together.

1047a14: ὥστε οὗτοι οἱ λόγοι ἐξαιροῦσι καὶ κίνησιν καὶ γένεσιν.

"And so these teachings brush aside movement as well as becoming."

Aristotle maintains that the Megarians do not acknowledge these

§ 21. Conclusion of the confrontation

phenomena in their full validity, and they do this not just in general but precisely there where they want to develop something in regard to δύναμις κατὰ κίνησιν and its ἐνέργεια, its being at work, and thus in regard to ἐνέργεια κατὰ κίνησιν. Aristotle's insight, then, receives its full weight only if one firmly grasps that this chapter continually deals precisely with δύναμις as ἀρχὴ κινήσεως, as δύναμις κατὰ κίνησιν. But to be sure, it does this in such a way that its direction of questioning necessarily brings the ἐνέργεια κατὰ κίνησιν into view.

Immediately following this, Aristotle explicitly elucidates through an example what it means not to want to see movement as an essential structural feature of δύναμις, and even to assert its essential non-being and thereby its foreignness to being.

1047a15-17: αἰεὶ γὰρ τό τε ἑστηκὸς ἑστήξεται καὶ τὸ καθήμενον καθεδεῖται· οὐ γὰρ ἀναστήσεται ἐὰν καθέζηται· ἀδύνατον γὰρ ἔσται ἀναστῆναι ὅ γε μὴ δύναται ἀναστῆναι.

"That is to say, both the standing will always remain standing and the sitting will remain sitting, that is, will not get up if it has sat down, since something is incapable of standing up if it does not have such a capability."

Considering all that has been said, this example needs no further explanation; between standing and sitting there are modes of *transition*, setting oneself down and standing oneself up. More exactly, these do not lie between the two as one stone is at hand between two others, but rather, sitting is having set oneself down, and standing, having stood oneself up. The transition belongs to the phenomena as that through which they must have gone or else will go, each in its differing way. Being capable of something is in its ownmost actuality co-determined through this phenomenon of transition. —It must be noted that this example is brought into the discussion anew along with the positive thematization of the guiding problem (line 26ff.), and without regard to what ensues with chapter six.

Aristotle has now made two things clear: (1) the untenable consequences of the Megarian thesis and (2) at the same time their failure to take up an orientation toward a central phenomenon—movement as transition and change. Now, in contrast, he shows what insight must be gained if both of these are to be avoided.

1047a17-20: εἰ οὖν μὴ ἐνδέχεται ταῦτα λέγειν, φανερὸν ὅτι δύναμις καὶ ἐνέργεια ἕτερόν ἐστιν· ἐκεῖνοι δ'οἱ λόγοι δύναμιν καὶ ἐνέργειαν ταὐτὸ ποιοῦσιν, διὸ καὶ οὐ μικρόν τι ζητοῦσιν ἀναιρεῖν.

"Now if these statements are not to be set forth, then it becomes apparent that capability and being at work (reciprocally) are different; and yet these statements allow both capability and being at work to be thrown together as one and the same thing, and thus they attempt to annul something that is not at all insignificant."

According to our partitioning of the entire chapter (cf. p. 150), this passage forms the conclusion of the first part, and thus the critical confrontation. As we now see, it contains as well the transition to the following second part, the positive solution to the guiding question. The Megarian thesis must collapse; this implies that the being present of the δύνασθαι qua δύνασθαι cannot be sought in enactment. If that happens, then ἐνέργεια is the actual δύναμις; both are one and the same, so much so that δύναμις as a potentially proper actuality disappears; it does not receive its due. The questioning concerning δύναμις qua δύναμις and not qua ἐνέργεια has no basis at all. If the Megarian thesis is thus relinquished, then, in any event, (at least) one thing is won: the view to the phenomena is not covered over by a violent theory; instead, one sees that being capable of something, and precisely thereby being at work, are in each case something different (ἕτερον). Accordingly, if ἐνέργεια is to be defined in the right way, then we must try in a reverse manner to save δύναμις and its way of being present in its proper essence, in order to put ἐνέργεια for its part into relief against this.

And so we surmise from the text quite unambiguously that this chapter depends precisely upon the elaboration of the heterogeneity of δύναμις over and against ἐνέργεια and vice versa. Thus there is a tangible progression in the inquiry. Of course, the usual interpretation could say, a progression perhaps, but of what kind? Now that, too, is decided quite unambiguously. In the passages which have just been

§ 22. *Actuality of being capable* 183

brought forth, Aristotle states twice that the Megarians brush aside and annul something and deny its value. First he states that κίνησις does not receive its due (a14); then: the difference between δύναμις and ἐνέργεια is not taken into account (a19-20). Evidently what is being annulled in both places hangs together. The brushing aside of the difference between δύναμις and ἐνέργεια is in itself the brushing aside of κίνησις. This can be the case only if just these two are essentially related to κίνησις, even while their difference must be observed. If κίνησις is rescued, then the difference between δύναμις and ἐνέργεια is secured as well.

With this, however, it is expressed as tangibly as possible: δύναμις and ἐνέργεια are here in chapter three taken as κατὰ κίνησιν. Not only is there no occasion to conclude that there is a premature introduction of the later theme of ἐνέργεια ἐπὶ πλέον, but it is stated with overwhelming clarity: here ἐνέργεια κατὰ κίνησιν is being dealt with. If what matters here is emphasizing the difference between ἐνέργεια κατὰ κίνησιν over and against δύναμις κατὰ κίνησιν, then this implies at the same time that the most proper theme besides δύναμις is the ἐνέργεια κατὰ κίνησιν.

All the same, we must now ask: (1) How is the preservation of the heterogeneity of δύναμις κατὰ κίνησιν and ἐνέργεια κατὰ κίνησιν connected with the securing of the phenomenon of κίνησις? (2) How is this securing for its part connected with the correct resolution of the guiding question of chapter three that we established; namely, how is it connected with the question concerning the being present of δύναμις qua δύναμις, of capability as capability prior to all actualization in enactment? The answer to this question must arise from an interpretation of the following positive discussion and determination of δυνατὸν ὄν ᾗ ὄν and of ἐνέργεια κατὰ κίνησιν.

§ 22. Ἐνέργεια κατὰ κίνησιν. *The actuality of being capable is co-determined by its essence—to this essence, moreover, belongs its actuality*

Before we continue with the interpretation, let us attempt briefly to coalesce still one more time the basic problem being dealt with here and to find its essential kernel. We can do this by taking up a difficulty which now suggests itself. On the one hand, the Megarians ought to be rejected on the basis of a contrary thesis: The actuality of δύνασθαι as such is not to be sought in ἐνεργεῖν. On the other hand, for a positive determination of the actuality of δύνασθαι as such, precisely ἐνέργεια ought now to come into play. How can both of these come together?

The obvious consequence of repudiating the Megarian thesis would be that then Aristotle as well would neglect for his part ἐνέργεια. And yet if we figure in this way, we presume that the Megarians, with their reference to ἐνεργεῖν as an explanation for the εἶναι of δύναμις qua δύναμις, also already possessed the correct insight into the essence of ἐνέργεια. But this is just what Aristotle contests.

That the Megarians relied upon ἐνεργεῖν does not at all prove that they had a proper notion of it. Just the opposite, they did not see precisely that ἐνέργεια qua ἐνέργεια is ἐνέργεια κατὰ κίνησιν. And they had to overlook this basic relationship because for them the view to the essence of κίνησις was in general distorted. But only if this essence also becomes clear does it become possible to comprehend δύναμις in its full content, and thereby to delimit the manner of its ownmost being actual. Δύναμις is indeed (according to chapters one and two) ἀρχὴ μεταβολῆς (or else κινήσεως)—that from out of which change and transition occurs. *How* something like this actually is can be determined only if it is continually being taken into account *what* this is. On the other hand, only through an adequate articulation of *how* δύναμις qua δύναμις actually is can *what* it is come to a full delimitation. And so the task of characterizing δυνατὸν ὄν ᾗ ὄν becomes at the same time the task of characterizing ἐνέργεια ᾗ ἐνέργεια, that is, the task of demonstrating that it is κατὰ κίνησιν and how it is so.

From this link found in the matter itself we can infer in advance that Aristotle for his part will not delimit δυνατὸν ὄν without reference to ἐνέργεια. First, however, this "not without reference to ἐνέργεια" implies by no means the identification of the actuality of capability with the actualization in enactment. And second, securing a suitable determination of the actuality of that which is capable as

§ 22. Actuality of being capable

such is to be found precisely in the correct formulation of the necessary relation of δύναμις to ἐνέργεια.

We shall divide the second portion of the whole discussion into individual sections as well.

1047a20-24: ὥστ' ἐνδέχεται δυνατὸν μέν τι εἶναι μὴ εἶναι δέ, καὶ δυνατὸν μὴ εἶναι εἶναι δέ, ὁμοίως δὲ καὶ ἐπὶ τῶν ἄλλων κατηγοριῶν δυνατὸν βαδίζειν ὂν μὴ βαδίζειν, καὶ μὴ βαδίζον δυνατὸν εἶναι βαδίζειν.

"So it could happen that something in fact is actual as something capable of something, and yet thereby not actually be that thing of which this actually capable thing as such is capable, and likewise it could happen that something capable is not actual as something capable, and yet is precisely and actually that of which it is capable; in the same way, this holds with regard to the other things that can be said about beings (κατηγορίαι in the most general sense); for example, that which is actually a being as a being capable of walking in actuality does not walk at all, and that which actually does not walk is nevertheless actually present as capable of walking."

If we examine the Greek text, we see εἶναι and μὴ εἶναι, being and non-being, starkly juxtaposed and contrasted. Both are even at the same time attributed to the same thing. If this, considered formally, is to be at all possible, then εἶναι and εἶναι must each be meant here in a different respect. This proves true. We have brought this out in the translation. In the preceding considerations we already expressed this in such a manner that we comprehended the being of δύναμις qua δύναμις as "actuality," and being in the sense of the actuality of that of which something capable is capable, as "actualization." The actualized has thereby actuality as well.

Aristotle introduces this thought with ὥστε: so it could happen, that is, if the heterogeneity between δύναμις and ἐνέργεια is taken into account. What does this imply for the resolution to the guiding question? To take into account the difference between δύναμις and ἐνέργεια means to attempt not to replace immediately the actuality of δύναμις with ἐνέργεια, thereby doing away with δύναμις. It means instead to attempt to see that δύναμις has its own actuality and to see how this is so. Aristotle initially secures this through a general statement which in turn is elucidated through an example: we come across an actual being which is capable of walking (δυνατὸν βαδίζειν ὄν); as something capable this is a being (ὄν), but it is a being which nevertheless is not yet or else is no longer walking.

Aristotle is satisfied with this reference. But such referring to a self-sufficient phenomenon is what is now decisive. And we may not forget that Aristotle already gave an important indication in his critical confrontation, according to which the right distinction between δύναμις and ἐνέργεια can occur only with the prior and consistent maintenance of κίνησις. But what does that mean? Nothing less than this: The being present of something capable as such and actuality in the sense of enactment are *modes of being in movement;* they are implicitly associated with this and are to be comprehended only on this basis.

Let us follow this indication. What do we gain for the clarification and determination of the actuality of something capable as such? Someone who is capable of walking, for example, but who does not enact this walking, how is such a capable one actual? Not walking, considered in terms of movement, is stillness, standing still. And yet is standing still so easily comprehended as the characteristic being at hand (Greek: presence) of something capable as such? Of course, in this case this is a necessary moment, but it alone does not suffice. Each one capable of going but not actually walking stands still and does not move. But such a person could in fact sit in a traveling ship and be just as actual as someone able to walk, even though he is in movement and not at rest. The fact that someone who is able to walk rests as such, this is meant evidently as such a way of moving, and this capable one is capable of this way of moving. The actuality of the capable is *co*-determined by a capable actuality, which shows up in enactment. It is co-determined in terms of such enactment; but it is not the same as such enactment.

How are we to comprehend this *co-determinateness*, that the enactment of capability in its own manner of actuality becomes visible in the actuality of something capable as such? Can we impress this upon ourselves through our own immediate experience? By all means.

§ 22. Actuality of being capable

Let us consider a sprinter who, for example, has (as we say) taken his or her mark in a hundred-meter race just before the start. What do we see? A human who is not in movement; a crouched stance; yet this could be said just as well or even more appropriately about an old peasant woman who is kneeling before a crucifix on a pathway; more appropriately, because with the sprinter we do not simply see a kneeling human not in movement; what we call "kneeling" here is not kneeling in the sense of having set oneself down; on the contrary, this pose is much more that of being already "off and running." The particularly relaxed positioning of the hands, with fingertips touching the ground, is almost already the thrust and the leaving behind of the place still held. Face and glance do not fall dreamily to the ground, nor do they wander from one thing to another; rather, they are tensely focused on the track ahead, so that it looks as though the entire stance is stretched taut toward what lies before it. No, it not only looks this way, it is so, and we see this immediately; it is decisive that this be attended to as well. What limps along afterwards and is attempted inadequately, or perhaps without seriousness, is the suitable clarification of the essence of the actuality of this being which is actual in this way.

What exhibits itself to us is not a human standing still, but rather a human poised for the start; the runner is poised in this way and is this utterly and totally. Thus we say—because we see it without looking any further—that he is poised for the start. The only thing needed is the call "go." Just this call and he is already off running, hitting his stride, that is, in enactment. But what does this say? Now everything of which he is capable is present [*anwesend*]; he runs and holds nothing back of which he would be capable; running, he executes his capability. This execution is not the brushing aside of the capability, not its disappearance, but rather the carrying out of that toward which the capability itself as a capability drives. The one who enacts is just that one who leaves nothing undone in relation to his capability, for whom there is now in the running actually nothing more of which he is capable. This, of course, is then the case only if the one who is capable comes to the running in full readiness, if in this readiness he extends himself fully. But this implies that he is then genuinely in a position to run only if he is in good condition, completely poised, in full readiness.

In a position to . . ., this means first: he is fit for it. Yet not simply this, but at the same time it also means: he ventures himself, has already become resolved. To actually be capable is the full preparedness of being in a position to, which lacks only the *releasement* into enactment, such that when this is at hand, when it has imposed itself, this means: when the one who is capable sets himself to work, then the enactment is truly *practice* and just this. It is nothing other than setting oneself to work—ἐνέργεια (ἔργον: the work or the product).

Now it becomes clearer how the actuality of δύνασθαι is to be comprehended through ἔχειν, having and holding, namely as holding oneself in readiness, holding the *capability itself in readiness*. This being held is its actual presence. In the example mentioned earlier, the potter who had lost both hands, the moment of passing beyond, of going over, is in a certain manner no longer at hand; the being held is no longer complete; the readiness is interrupted.

If you have followed this entire clarification of the essence of that which is capable and actually present with a continual view toward the phenomenon (the runner immediately before the start), then the "definition" which Aristotle now gives for δυνατὸν εἶναι may no longer be foreign to you.

> 1047a24-26: ἔστι δὲ δυνατὸν τοῦτο, ᾧ ἐὰν ὑπάρξῃ ἡ ἐνέργεια οὗ λέγεται ἔχειν τὴν δύναμιν, οὐδὲν ἔσται ἀδύνατον.

"That which is in actuality capable, however, is that for which nothing more is unattainable once it sets itself to work as that for which it is claimed to be well equipped."

Here we have again one of the unprecedented and determining essential insights, through which Aristotle for the first time illuminates a previously obscure realm. In this concise statement, every word is significant. With Aristotle the greatest philosophical knowledge of antiquity is expressed, a knowledge which even today remains unappreciated and misunderstood in philosophy.

§ 22. Actuality of being capable

*   *

*

What is required now is by no means more tedious repetition for the purpose of bringing closer to you how the entire preceding inquiry works toward this statement, so that it must, as it were, spring forth. Only a few aspects of this "definition" are to be indicated. The first one is the very first word: ἔστι. This may not simply be taken as "is" in the sense of a what-being, so that we would be able to translate: being capable is that for which. . . . In accord with the overriding theme of the chapter, what is being dealt with here is indeed not what we have to understand under being *capable;* that is said in chapters one and two. Instead, the task is to determine that in which the *being* of something capable, its actuality—the εἶναι of the immediately following sentence—consists. But this is—to my knowledge—completely missed in all the interpretations and translations; every prospect for an understanding of the definition is thereby eliminated from the very beginning.

The second thing that has to be noted is that Aristotle does not simply speak about ἐνέργεια, being at work, putting or setting oneself to work, but rather quite unmistakably about that very setting oneself to work for which the capability under discussion is well equipped. An essential difference over and against the Megarian thesis is hereby expressed. e see that Aristotle also draws ἐνέργεια into the delimitation of the actuality of the δυνατόν qua δυνατόν, but not in the general sense of an enactment emerging from nothing. Instead he includes it in its ever-determinate relatedness to its respective capability.

The third thing is the correct understanding of the ᾧ οὐδὲν ἔσται ἀδύνατον. This must be understood so that it remains related to ἐνέργεια. This means: that which is fully and actually in-a-position-to is just that present being which in enactment must leave *nothing unattained.*

It actually would not be worthwhile to go into all the empty cleverness of the interpreters who, with a certain disguised feeling of superiority, think they have finally caught the great Aristotle here making a capital error. But please, does not every child see what kind of famous definition Aristotle offers us? Δυνατόν—ὃ οὐκ ἀδύνατον: potential is that which is not non-potential. And yet with this wisdom the most impossible becomes possible, which almost seems to be the rule in the usual interpretations of philosophy. I have to leave it up to you to refute these overly clever pedantries.

Only a few hints, whereby what has already been said is basically repeated: (1) The definition of what is potential is not being dealt with at all here, but rather the definition of what is capable, which is not the same thing. (2) Neither is the definition of something capable with regard to its what-content being dealt with, but instead with regard to the actuality which is essentially proper to it. (3) Nowhere is it simply stated: δυνατόν = τὸ οὐκ ἀδύνατον. Instead, if οὐκ ἀδύνατον is said of δυνατόν, then this is so only when οὐκ ἀδύνατον is made to fulfill the condition of ἐὰν ὑπάρξῃ ἡ ἐνέργεια. Thus the "not incapable" is not simply attributed to the "capable," but rather, if it is this, then this pertains to its actuality. But if one already wants to retreat to the vacuous assurance of academic logic, according to which with a definition the definiendum may not enter the definitum, then it must be objected precisely what type of definition do we have here before us: whether it holds when defining a table, chair, house, ox, or donkey, or whether it pertains to that which lies far from all such things, so far that even today it remains out of reach of all pondering cleverness.

What has been said should lead one to the insight that it does not help at all if we think through this definition by means of a purely abstract deliberation. In this way we remain blind; we do not see what is being discussed, nor do we see how Aristotle with unprecedented certainty brought this to word from out of that which offers itself to the truly philosophizing vision.

Aristotle includes an illustration with the definition; the examples can serve to demonstrate the delimited essence in various manifestations.

1047a26-29: λέγω δ' οἷον, εἰ δυνατὸν καθῆσθαι καὶ ἐνδέχεται

§ 22. *Actuality of being capable* 191

καθῆσθαι, τούτῳ ἐὰν ὑπάρξῃ τὸ καθῆσθαι, οὐδὲν ἔσται ἀδύνατον· καὶ εἰ κινηθῆναί τι ἢ κινῆσαι ἢ στῆναι ἢ στῆσαι ἢ εἶναι ἢ γίγνεσθαι, ἢ μὴ εἶναι ἢ μὴ γίγνεσθαι, ὁμοίως.

"I understand this, however, in this way: If one is capable of sitting in such a way that he can allow himself to sit, then nothing will remain unattained when it comes to sitting. The same holds when something is capable of being moved or of moving, of standing or of bringing something to stand, of being or of becoming, of not being or of not becoming."

For our interpretation it is worth noting the καὶ ἐνδέχεται which is here linked to δυνατὸν ἐστιν; this is so because δύνασθαι and ἐνδέχεσθαι are now used interchangeably. I have translated the ἐνδέχεσθαι here according to what we found to be the characteristic determination of being actually capable: to be fully in readiness, to be able to take something upon oneself. The καὶ is to be taken here as an expository "and"—"and even in the manner that . . . " Moreover, we encounter ἐνδέχεσθαι again in chapter eight of the treatise. The connection to ἐνδέχεσθαι is further proof that the question concerns the actuality of capability rather than what capability is.

But on the other hand, we now see precisely: what capability is, namely ἀρχὴ κινήσεως—related to change and therefore enactment; this co-determines the manner and mode of actuality of δύνασθαι, the essence of its being present. Thus we are now able to say: The question concerning the essence of δύναμις is thereby first thoroughly posed and resolved when the essence of the accompanying actuality is also determined along with it.

In philosophy, and even quite often through an appeal to antiquity, the *question of essence* is understood generally in such a way that what is at issue in this question is what something is, its what-being, without regard to whether it is actual or not. Actuality is here irrelevant. But this is ambiguous—and philosophy has succumbed to this ambiguity. It has mostly neglected to ask what then is the essence of actuality. And when the question is posed, this occurs in such a way that actuality, *existentia*, is taken in a broad, all-encompassing sense; the actual is then what is present, at hand. It is not seen that this very actuality is essentially transformed with the essence in the more narrow sense, where only what-being is expressed. The full essence of a being, however, and this is something which we first have to learn to understand, pertains both to the what of a being and to the how of its potential or actual actuality. Of course, what a thing is must be determined without regard to whether it is actual or not; the essential determination of a table holds also for a potential table or a table which is no longer at hand. And yet not to consider whether the what-being is actual or not does not at all mean that it also matters little whether it is asked how this actuality according to its essence is, an actuality which is prescribed for this respectively determined what-being.

Admittedly, Aristotle did not in our context explicitly unfold the question of a full knowledge of essence. Although he did, in fact, bring the delimitation of the essence of actuality into the closest discerning connection with the determination of what a capability is. But for reasons which lie locked in the ancient and Western conception of being and thereby of what-being, neither is this central problem of the question of essence posed later.

We contend that the guiding question relating to the actuality of δύναμις κατὰ κίνησιν compels one to take ἐνέργεια κατὰ κίνησιν also into regard, and to draw it as well into the definition of this actuality, not as is the case with the Megarians but precisely in such a way that the relation of ἐνέργεια to κίνησις and thereby to δύναμις becomes apparent. Only when ἐνέργεια is necessarily thematic in this manner does it now make sense for Aristotle to begin to speak explicitly about the word and the word's significance. The way this occurs must lay aside completely all doubts about the theme of chapter three.

1047a30-32: ἐλήλυθε δ' ἡ ἐνέργεια τοὔνομα, ἡ πρὸς τὴν ἐντελέχειαν συντιθεμένη, καὶ ἐπὶ τὰ ἄλλα ἐκ τῶν κινήσεων μάλιστα· δοκεῖ γὰρ [ἡ] ἐνέργεια μάλιστα ἡ κίνησις εἶναι.

"It is, however, the name and meaning of ἐνέργεια—being at work, a meaning which in itself is directed toward ἐντελέχεια—holding itself in completion, which has also gone over to the other being, namely from its prevailing usage in reference to movements; for mostly and

§ 22. Actuality of being capable

primarily movement shows itself as that whereby something is 'in process,' at work, in full swing."

(Something twofold is stated: (1) the connection with ἐντελέχεια; (2) the matter of ἐπὶ τὰ ἄλλα, whereby not ἐπὶ πλέον is meant, but rather that which is spoken about in the previous sentence. Enacting is related not only to being able to move and movement, but instead the essence of enactment is being at work, setting oneself to work. Ἐντελέχεια refers to this as well; compare what was said earlier about ἔργον and τέλος: ἐντελέχεια: the end, to possess completion as that which has been carried out, to hold oneself in it—properly: being produced.)

Editor's Epilogue

The announcement for Heidegger's 1931 summer semester course at the University of Freiburg read: "Interpretations from ancient philosophy; Tu–Th 5-6." Heidegger began the course on April 28, and it ended on July 30. His manuscript "*Interpretations of Ancient Philosophy*/Aristotle, *Metaphysics* Θ" consists of folio pages written in crosswise format, with the running text written exclusively on the left half, and the insertions, corrections, extensions, and additions written on the right half. The pagination with various subdivisions runs to page 47; there are in actuality 56 pages. In addition, there are at least half as many supplements and annotations, especially with the notices for the recapitulations. The presentations of the course ended on the bottom of p. 189 above.

The manuscript was edited according to the guidelines of Martin Heidegger, as implemented in the lectures that he himself edited. Chiefly, the entire manuscript was made into a complete and book-ready transcript. The editor then divided the text into four parts, and provided the accompanying headings and titles. An asterisk marks the end of each lecture period; it is followed by a one-paragraph recapitulation (except for the last two lectures, where no recapitulation is given). Heidegger's comments on and corrections of the Aristotle text are inserted in brackets and designated as his; by contrast, the various addenda in his translation appear simply in parentheses. A prior copy of the manuscript edited by H. Feick (not decisively arranged and classified) was extremely helpful for deciphering and corroborating terms. The same is true for two exceedingly judicious lecture transcripts (one shorter, the other more ample), which were consulted to help resolve problems with the recapitulations, the completion of the translation, and occasionally in rounding off and securing the chain of thought.

The text of the *Metaphysics* that is cited is *Aristoteles' Metaphysik*, recognovit W. Christ (Leipzig, 1886; nova impressio correctior 1895 and later), reprinted in *Aristoteles' Metaphysik,* Greek and German, translated by H. Bonitz, newly prepared with introduction and commentary by H. Seidl (2 volumes; Philos. Bibl. 307 and 308, Hamburg, 1978/80). Further aids are *Aristotelis Metaphysica*, recognovit et enarravit H. Bonitz, volume 1 (text) and volume 2, *Commentarius* (Bonn, 1848/49; reprint of the commentary, Hildesheim, 1960); *Die Metaphysik des Aristoteles*, basic text, translation and commentary together with explanatory essays by A. Schwegler (4 volumes; Tübingen, 1847/48; reprinted in two volumes, Frankfurt am Main, 1968); Aristotle's *Metaphysics*, a revised text with introduction and commentary by W. D. Ross (2 volumes; Oxford, 1924; corrected editions 1953 and later); W. Jaeger, *Aristoteles. Grundlegung einer Geschichte seiner Entwicklung* (Berlin, 1923; 3rd ed., Dublin/Zürich 1967).

The lecture course now available attests to the search for a connecting horizon for the encounter with that which was thought in advance by Aristotle. The Introduction outlines a basic sketch of Aristotle's philosophy in general. Within this belongs the question of *dynamis* and *energeia*, which Aristotle discusses in Book IX of the *Metaphysics*. The line-by-line interpretation of the first three chapters of this book deals with the essence and actuality of force. The phenomenon of force or capability, which is discussed thoroughly in its variations, becomes the nucleus for splitting up the general Greek being-concept of presence, and also becomes a guiding thread for Heidegger's determination of truth, as it emerges in the concept of being. Thus, this text provides access for understanding the connection of the whole of Book IX to the final chapter, and also prepares for a broader and farther-reaching discernment of Greek philosophy. This is the sense in which the original title is to be understood. However, a narrower and more focused title for the present volume appeared advisable.

I wish to express my heartfelt thanks to Professor Klaus Held. His unconditional support and encouragement made this publication possible. For their advice on questions regarding the form of the text, I am indebted to Dr. Hermann Heidegger and Dr. Friedrich-Wilhelm v. Hermann.

H. Huni
Wuppertal, November 1979

Glossary of German Words

Aneignen: acquire, acquisition
Ankündigung: announcing
Anlage: proficiency
Anwesenheit: presence
Aufhören: cease
Aus-der-Übung-sein: being out of practice
Auseinandersetzung: confrontation
Ausgang: origin
Aushaltsamkeit: endurance
Ausrichtung: orientation
Aussehen: aspect, outward appearance
Ausübung: practice, practicing
Befähigung: competence
Begabung: talent
Bekunden: witness
Bereich: realm
Bewegtes: beings that move, being-moved
Beziehung: relationship
Bezug: relation
Bildsamkeit: malleability
Brüchigkeit: fragility
Dichtung: poetry
Durchhalten: put up with, come through
Eigentümlichkeit: peculiarity
Eignung: aptitude
Einbezug: implication
Eingeübtsein: being trained
Einübung: training
Entgegenliegendes: contrary
Entzug: withdrawal
Erdulden: endure

Erkunden: explore
Erleiden: tolerate
Ertragen: bear
Ertragsamkeit: bearance
Fähigkeit: capacity
Fügung: jointure
Gefüge: structure
Gegenteil: contrary
Geschicklichkeit: skill
Gewalt: violence, violent force
Haben: having
Hergestelltheit: having been produced
Herstellen: produce
Können: ability
Kraft: force, power
Kräftigsein: being powerful
Kraftsein: being a force
Künden: declare
Kundgeben: give notice
Kundige Kraft: conversant force
Kundigsein: being conversant
Kundmachen: make known
Kundnahme: take notice
Kundschaft: conversance
Kunst: art
Leiden: suffer
Leitbedeutung: guiding meaning
Macht: power
Namensgleichheit: nominal identity
Offenbarkeit: manifestness
Potenz: power
Sichverstehen auf: versatile understanding of
Streben: strive
Tätigkeit: activity
Übertragen: metaphorical, transfer
Umgehen: comport

Umschlag: change
Unkraft: unforce
Unkräftig: powerless, forceless
Verkünden: proclaim
Vermögen: capability
Verwirklichung: actualization
Vollzug: enactment
Von-wo-aus: from out of which
Vorhanden: present, at hand
Vorhandensein: being present
Wahrnehmen: perceive
Weggeben: give up
Wesen: essence
Widerständigkeit: resistance
Wirken: effect
Wirklichkeit: actuality
Zerbrechlichkeit: breakability
Zugehörigkeit: belonging(ness)
Zwiespältigkeit: divisiveness

Glossary of Greek Words

ἁπλῶς: purely and simply (*einfachhin*), simply (*einfach*)
ἀδυνατόν: powerless, forceless (*unkräftig*)
ἀκολουθεῖν: follow, constantly going after, always already going along with
ἀληθές: uncovering
ἀναλογία: analogy (*Analogie*)
ἀναλέγειν: correspond (*Entsprechen*)
ἀπόφανσις: assertion
ἀπόφασις: affirmation
ἀρχή: origin (*Ausgang, Von-wo-aus*)
ἄπειρον: the unbounded
αἴσθησις: perceive, take for true (*Wahrnehmen*)
βίος: life, life history
γένος: genus (*Gattung*)
δύναμις: *potentia,* force (*Kraft*), capability (*Vermögen*), possibility (*Möglichkeit*)
δύναμις ἕξις ἀπαθείας: force of resistance
δύναμις μετὰ λόγου: capability (*Vermögen*), conversant force (*kundige Kraft*)
δύναμις τοῦ παθεῖν: force of bearing
δύναμις τοῦ ποιεῖν: force of doing, producing
δυνατόν: powerful, forceful (*kräftig*)
εἶδος: aspect, outward appearance (*Aussehen*)
ἔμψυχον: besouled
ἐν ἄλλῳ ἤ ᾗ ἄλλο: in another or to the extent that it is another
ἐναντίον: contrary
ἐνέργεια: *actus,* actualization, actuality, being at work
ἐντελεχέια: holding in completion
ἐπὶ πλέον: extending further
ἐπιστήμη: science, familiarity with things
ἔργον: work (*Werk*)

ἑρμηνεία: interpretation (*Auslegen*)
ἔχειν: have, possess
φρόνησις: circumspection
κίνησις: movement (*Bewegung*)
καθ' αὑτό: in respect to self, self-same
κατὰ κίνησιν: with regard to movement
κατὰ συμβεβηκός: with respect to being co-present
κατηγορία: category (*Kategorie*), that saying which is involved in every assertion in a preeminent way
κινούμενον: moved being (*Bewegtes*)
λήθη: forgetting
λόγος: discourse, conversance
λέγειν: gathering, to gather, bring into relation
λέγεται πολλαχῶς: said in many ways, understood in manifold ways
μὴ ἔχειν: not having
μεταβολή: change (*Umschlag*)
ὁμώνυμον: nominally identical (*namensgleich*)
ὀρεκτόν: what is striven after
ὄρεξις: striving (*Streben*)
ὁρισμός: delimitation (*Umgrenzung*)
παρουσία: presence
πέρας: boundary, limit
ποίησίς: production (*Herstellung*)
ποιεῖν: produce, bring forth
ποιητικὴ ἐπιστήμη: versatile understanding of ποίησίς
ποιόν: being so constituted (*Beschaffensein*)
πολλαχῶς λεγόμενον: what is said in many ways
πολλαχῶς: manifold, in many ways
στέρησις: withdrawal (*Entzug*)
τέχνη: capability for producing
τὸ εἶναι: being (*Sein*)
τὸ ὄν ᾗ ὄν: beings as such (*das Seiende als solches*)
τὸ ὄν: beings (*das Seiende, Seiend*)
ὕλη: that out of which something is to be produced
ψεῦδος: distortion, concealment

ψευδής: deceptive, concealing
ζῷον λόγον ἔχον: the living being that has λόγος
ζωή: life, living being, animal